THE VITAMINS

Their Role in Medical Practice

THE VITAMINS

Their Role in Medical Practice

by

John Marks, MA, MD, FRCP, FRCPath

Fellow, Tutor and Director of
Medical Studies
Girton College, Cambridge

MTP PRESS LIMITED
a member of the KLUWER ACADEMIC PUBLISHERS GROUP
LANCASTER / BOSTON / THE HAGUE / DORDRECHT

Published in the UK and Europe by
MTP Press Limited
Falcon House
Lancaster, England

British Library Catologuing in Publication Data

Marks, John
    The vitamins: their role in medical practice.
    1. Vitamins in human nutrition
    I. Title
    613.2'8   TX553.V5

ISBN 0-85200-851-1

Published in the USA by
MTP Press
A division of Kluwer Boston Inc
190 Old Derby Street
Hingham, MA 02043, USA

Library of Congress Cataloging in Publication Data

Marks, John, 1924 –
    The vitamins: their role in medical practice.
    Includes bibliographies and index.
    1. Avitaminosis – Handbooks, manuals, etc.
    2. Vitamins – Handbooks, manuals, etc. I. Title.
    [DNLM: 1. Vitamins.   QU 160 M346v]
    RC623.7.M37 1985   612'.399   85 - 17154

ISBN-13: 978-94-011-7323-0   e-ISBN-13: 978-94-011-7321-6
DOI: 10.1007/978-94-011-7321-6

Phototypesetting by Vantage Photosetting Co. Ltd.
Eastleigh and Southampton

Contents

CONTENTS

Preface

This book has been designed, as its title implies, as a practical book for medical practitioners, although it should be of interest to medical students and nutritionists. It attempts to provide essential information about this important group of substances rather than be an all embracing monograph on the subject. For this reason biochemical and physiological considerations have been kept to a minimum, and aspects of animal disorders and animal husbandry have not been considered. The material is often presented in a rather dogmatic fashion and, with rare exceptions, references are not included since this makes reading more difficult.

The exceptions, where references are provided, are the therapeutic claims, and the series of recent studies which have indicated that vitamin deficiencies are still widely present among certain groups of the population of many industrially developed countries. To add to this reference list there is a reading list which has been selected to give key books, reviews with extensive bibliography and important articles over the past 10 years. From this reading list it is possible to trace most of the literature on the vitamins since they were first described over half a century ago.

John Marks
Girton College, Cambridge

Introduction

The vitamins consist of a mixed group of chemical substances, designated as vitamins not by their chemical characteristics but by their function. They are organic constituents of the diet, essential for the well-being of the animal body and ingested in small amounts. However, the application of the definition is somewhat arbitrary, for some organic compounds, e.g. the essential fatty acids are not usually regarded as vitamins, while some vitamins, e.g. vitamin D, are predominantly synthesized by the body and could be better regarded as pro-hormones. Hence the definition stems from historical rather than current considerations.

The vitamins may be classed into fat soluble and water soluble groups, and this arbitrary but useful classification has remained the standard. At the time of their discovery as essential food constituents their chemical identity was unknown, and in consequence a system of designation by letters developed with quantities defined in units, usually derived from animal curative tests. However, the chemical nature of the active compounds is now known, and generic (trivial) names have been established. For several of the vitamins alternative chemical forms may show biological activity, often by conversion to the active tissue metabolite. Hence the trivial name may represent *an* active form and not *the* active form. The trivial name cannot, therefore, be regarded as the true synonym of the biological designation though it is often used in this way. Table 1 lists the alphabetical designations, the most accepted alternative names and the 'trivial names'.

It will be seen from Table 1 that many (but not all) of the water soluble vitamins were given an alphabetical designation of vitamin B with a numerical suffix. Several of these water soluble vitamins are now usually referred to as the 'B group'. This group comprises thiamine, riboflavine, niacin, pyridoxine and pantothenic acid. Cobalamin and folic acid are also usually included in the group and biotin is also sometimes added. Strictly speaking all these eight substances should be included.

1

Table 1 Designation of currently accepted 'vitamins' according to common usage, trivial names of the main form(s) and alternative names (those that are now largely obsolete are shown in square brackets). The problem of these names not being true synonyms is described in the text

Common usage name	Trivial name	Alternative names (representative)
Fat soluble		
Vitamin A	retinol	axerophol, [antixerophthalmic vitamin]
Vitamin D	calciferol	[antirachitic vitamin]
Vitamin E	α-tocopherol	[antisterility vitamin]
Vitamin K	phylloquinone	phytylmenaquinone
Water soluble		
Vitamin B_1	thiamin(e)	aneurine, [antineuritic vitamin]
Vitamin B_2	riboflavine	[lactoflavine]
Niacin	nicotinamide	vitamin PP, vitamin B_3, nicotinic acid
Vitamin B_6	pyridoxine	as pyridoxol, pyridoxal or pyridoxamine
Vitamin B_{12}	cobalamin	[antipernicious anaemia factor]
Folic acid	folacin	[vitamin M or B_C or B_9, lactobacillus casei factor]
Pantothenic acid	pantothenic acid	[vitamin B_5]
Biotin	biotin	[vitamin H]
Vitamin C	ascorbic acid	[antiscorbutic factor]

In addition to these 13 biologically active current members of the vitamin class many other substances have in the past been designated vitamins. Some are not now thought to be physiologically active, others though accepted as active are no longer termed vitamins. The list of obsolete

Table 2 Designation of 'obsolete vitamins' according to their alphabetical terminology and trivial (generic) names

Vitamin	Trivial name	Alternative designation
B_4	adenine	
B_7	choline	vitamin J
B_{10}	p-amino benzoic acid	
B_{11}	carnitine	vitamin 0
B_{13}	orotic acid	
B_{14}	xanthopterine	
B_{15}	pangamic acid	
B_{17}	laetrile	
C_2	rutine	vitamin P; flavinoids
F	linoleic acid	essential fatty acids
I	meso-inositol	
N	lipoic acid	

vitamin designations is given in Table 2, since the vitamin designation of these substances is still occasionally encountered in practice. This group of substances is sometimes designated vitaminoids but this term should be avoided since it is also used for substances which are chemically related to the vitamins that are currently used therapeutically.

For most vitamins the daily requirement is expressed now, not in arbitrary units but in metric weights. There will be, however, a few exceptions to this, for some of the substances (particularly those where alternative active chemical forms exist) are still more commonly known by the older terms. Hence when there is some doubt we have tended to give the international unit designation.

The biochemical function of several of the vitamins is now known. Thus for example, cholecalciferol is a pro-hormone in calcium metabolism, while the members of the B group are components of coenzymes essential for

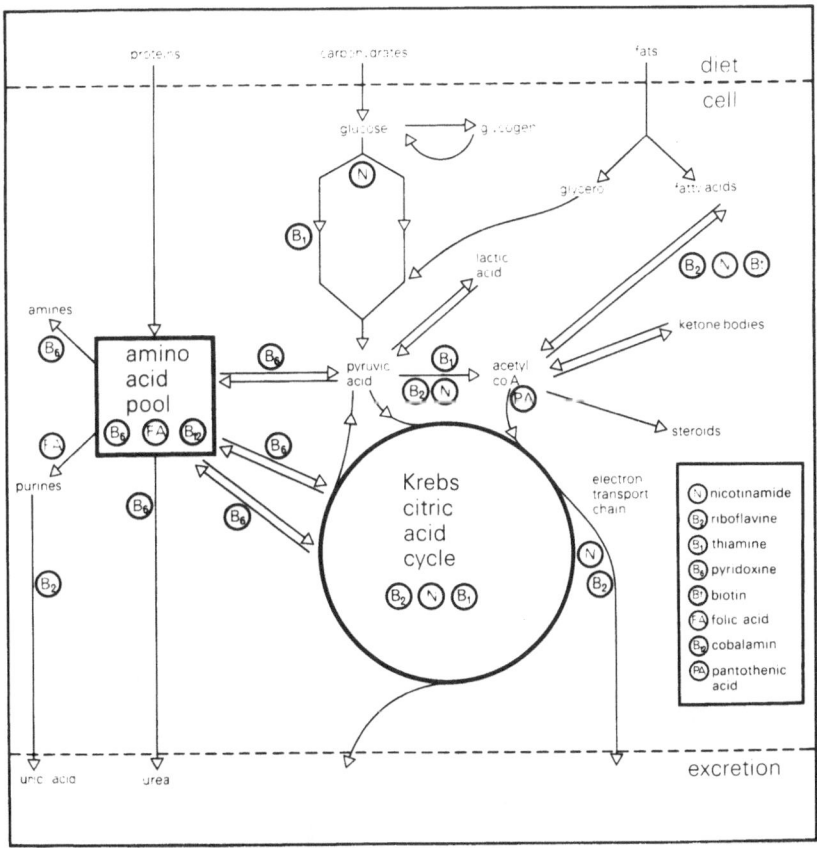

Figure 1 The importance of the vitamins of the B complex in cellular metabolism. The main activities are shown in diagrammatic form

Table 3 Summary of the vitamins, their storage, active forms and principal metabolic functions

Vitamin	Storage form	Active forms	Principal metabolic functions
Retinol	retinyl fatty acid esters largely protein bound in the liver	(a) ll-cis-retinal (b) retinoic acid	(a) vision, as rhodopsin (b) metabolism in cells of ectodermal origin
Calciferol	calciferol	1,25-dihydroxychole-calciferol	via calcium binding protein, regulation of calcium and related substances, interacting with parathyroid hormone and calcitonin
Tocopherol	tocopherol esters	tocopherol	(a) physiological antioxidant stabilizing various membranes and tissue components, particularly lipids (b) role in intracellular respiration
Phylloquinone		(a) ? phylloquinone (b) menaquinones-4,-5 and -6	(a) role in the formation of various clotting factors (b) electron transport in respiratory chain
Thiamine	small amount as TPP	thiamine pyrophosphate (TPP)	(a) coenzyme of pyruvate decarboxylase – decarboxylation (b) coenzyme of 2-oxo-glutarate-dehydrogenase – oxidation (c) coenzyme of transketolase
Riboflavine		(a) riboflavine-5'-phosphate (FMN) (b) flavine-adenine-dinucleotide (FAD)	both these active forms (FMN and FAD) form the prosthetic groups of a large number of enzymes which act as hydrogen transfer agents particularly in the metabolism of fatty acids and amino acids
Pyridoxine		pyridoxal-5'-phosphate	(a) coenzyme of several enzymes which play a vital role in protein metabolism (b) synthesis of biogenic amines for brain activity

4

Niacin		(a) nicotinamide-dinucleotide (NAD) (b) nicotinamide-dinucleotide-phosphate (NADPH)	(a) coenzymes of various dehydrogenases (preferentially exerted by NADPH) (b) Specifically in transfer of hydrogen ions in intermediary metabolism mainly to flavo-enzymes
Pantothenic acid		coenzyme A (CoA)	key role in acyl transfer systems in both anabolism (e.g. fatty acids, steroids, etc.) and in catabolism (e.g. in citric acid cycle, oxidative breakdown of fatty acids)
Folic acid	small stores as folic acid in liver	tetrahydrofolic acid	transfer of labile 'C_1 groups' in, e.g. amino acid metabolism and nucleotide production
Cobalamin	cobalamin in liver	mainly as 5-deoxyadenosyl-cobalamin	transfer of labile 'methyl groups' with folic acid in transmethylation particularly in nucleoprotein metabolism
Biotin		1'-N-carboxybiotin	prosthetic group of carboxylating enzymes which take part in synthesis of fatty acids, in gluconeogenesis and amino acid catabolism
Ascorbic acid	ascorbic acid in various sites	ascorbic acid – dehydro-ascorbic acid redox system	essential redox system for reversible transfer of hydrogen ions and electrons. In hydroxylation systems, particularly for corticosteroid and catecholamine synthesis

cellular metabolic pathways (Table 3). Details of the chemical reactions and physiological functions are considered under the individual vitamins, but a simplified version of the many functions of the vitamins in the complex biochemistry of the living cell is given in Figure 1.

Individual vitamin requirements for the maintenance of health are now known with reasonable accuracy, though there are some notable trans-national differences of opinion. Considerable differences of opinion still exist among nutritionists and physicians concerning the adequacy of the vitamin content of the average diet. There is a general feeling that the diet of people living in most industrialized countries is adequate for the supply of all nutrients, including vitamins. On the other hand, several recent surveys have shown that specific groups of 'normal people' have reduced vitamin reserves, clear biochemical and often some clinical evidence of vitamin deficiency. The differences in opinion stem in large measure from the definitions of 'average diet' and 'normal people'. Typical well-balanced diets as described in nutrition handbooks do indeed provide adequate amounts of the vitamins but it is necessary to appreciate that even in an affluent society deviation from a well-balanced diet is not uncommon.

In animal experiments it is possible to demonstrate improved perfor-mance from dietary vitamin levels in excess of the minimum required for apparent health. The concept of optimum levels, as opposed to adequate dietary intakes, is now well established for animals, but the currently available data do not allow us to define similar optimum levels for the human. Nevertheless, there would seem to be a need for further investigation of this problem, and in particular the effects of common stresses (e.g. infections, physical activity, rapid growth, etc.) on vitamin needs. A discussion of the effects of different dietary levels of some of the vitamins in humans is found on page 21.

One of the main problems for the assessment of the appropriate dietary level for any vitamin in the human diet is the determination of the significance of the different criteria used. Thus if there is a vitamin deficiency in the diet the consecutive states that follow in the body are (Figure 2):

reduced vitamin reserves,
deterioration in biochemical function,
initial clinical evidence of disease – then, in many cases,
irreversible histopathological changes.

Currently dietary adequacy is usually based on the prevention of clinical signs of deficiency, usually with some relatively minor safety margin. It could be logically argued, however, that an adequate diet is one that maintains adequate tissue reserves. Several authorities now believe that if such levels, which correspond to the optimum levels accepted now for animals, were maintained in the human, they would result in increased performance and reduced incidence of disease.

6

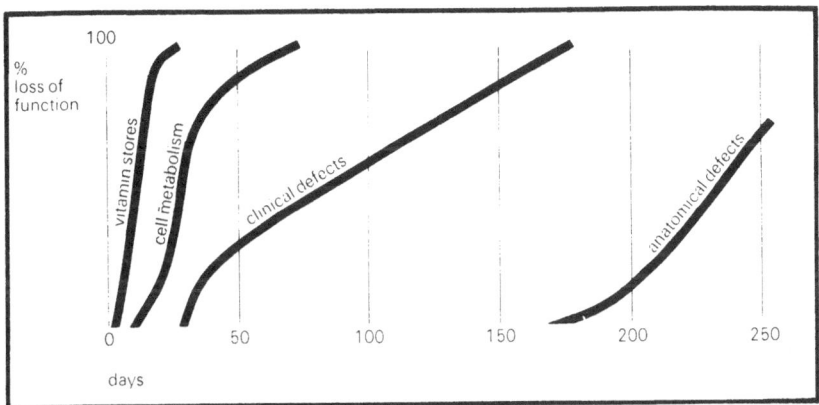

Figure 2 A schematic representation of the development of a vitamin deficiency

Nor for the majority of the vitamins is there evidence of derangement from high levels of intake. Indeed pathological signs of hypervitaminosis have been reported so far for very few of the vitamins despite reports of levels of intake greatly in excess of that recommended for dietary adequacy. Hence with very few exceptions the vitamins may be regarded as very safe compounds, in comparison to many ingested chemicals.

Minor changes in chemical structure of the vitamins may not only destroy the primary vitamin activity but produce more pronounced alternative effects or produce a direct antagonism to the parent compound. The vitamin antagonist may produce its effect either by interfering with the conversion of the vitamin to its coenzyme form or by displacing the coenzyme from its combination with the protein part of the enzyme. However, the effect of administering an antagonist may not mimic the effects of the dietary deficiency, particularly if several different enzyme systems utilize the same coenzyme. The order and extent to which the enzyme systems are affected depends in large measure on the binding potential of the antagonist at different enzyme sites. Vitamin antagonists have been used to study vitamin deficiencies, but the results should be interpreted with caution. They have also been used for therapeutic purposes, to influence cellular metabolism, particularly in the cell over-growth of malignant disease. Though theoretically attractive, the results have been disappointing in practice, with the exception of the use of anti-folate agents in childhood leukaemia (page 174) and vitamin K antagonists as anticoagulants (page 140).

More recently it has been recognized that chemical modification of certain of the vitamins produces an enhanced level of some of the activities of the natural substance – though often with increased toxicity levels as well. Some of these effects are just reaching the therapeutic application stage (page 120), though in most circumstances the level of toxicity which

7

exists with the current active substances means that they can only be advised for specialist hospital use.

As explained above, the majority of clinical manifestations of vitamin deficiencies are the direct result of an inadequate dietary intake, but rarely they may be due to an hereditary deficiency of the relevant enzyme. These include among others, the pyridoxine dependency state (page 157) and Hartnup disease, a genetic variant of pellagra (page 161). Other diseases and other therapies can also influence vitamin needs.

Thus the current story of the vitamins is one of an awakening to an understanding that these substances have an important part to play, not only in the maintenance of health, but in the therapy of disease.

We have now come out of the period of the vitamin doldrums, when early unsubstantiated therapeutic claims led to vitamin therapeutic nihilism. This book tries to provide the busy family physician with a working outline of our present views about the importance of these substances in medical care today. It has been divided into three distinct parts: general considerations about the vitamins; aspects of direct practical relevance to the busy practising physician and a more detailed description of each of the individual vitamins.

PART 1

GENERAL CONSIDERATIONS

1

Vitamin needs

There is no clear and definite definition for the vitamin needs of the individual, nor indeed for the vitamin needs of a group. This stems in part from the fact that different people have different requirements depending upon variations of, e.g. their activity and other dietary constituents.

However, apart from this interindividual variation there is no clear definition of what exactly constitutes an adequate intake for the human. Thus, for example, some authorities have regarded as appropriate a level of intake which will prevent, with a reasonable margin, the development of physical signs of a vitamin deficiency. Other authorities have held that, since the clinical signs of a vitamin deficiency occur very late in the slide from vitamin adequacy through metabolic defects to clinical ill health (see Figure 2), an adequate diet should contain sufficient of each of the vitamins to maintain tissue metabolic integrity. If this is regarded as the appropriate definition then it follows that the requirements are clearly higher than those for the prevention of clinical disorders. Still other authorities have considered that where there is evidence of vitamin storage then an adequate intake should supply sufficient vitamins to maintain the vitamin stores at an appropriate level.

In animals it is known that levels of vitamins in excess of those required purely for the maintenance of health lead to improved performance. The evidence for this in humans is less satisfactory, but the general concept is nevertheless accepted by many under the term 'optimum nutrition'. This aspect will be considered later (page 21).

Factors that influence the needs

Exercise

Exercise influences the requirements due to the use of carbohydrates as metabolic fuels, and possibly also due to the loss of vitamins in the sweat,

Table 4 Recommended dietary allowance for young men under various conditions of physical activity. Differences are demonstrated by reference to values from four representative countries and are shown as percentage variations from the value for moderate activity

Vitamin	Activity state	UK	West Germany	USSR	USA
Thiamine	sedentary	1.0 mg − 17%	1.7 mg − 23%	1.6−1.7 mg	*1.2 mg − 14%
	moderately active	1.2 mg	2.2 mg	1.7−1.8 mg	1.4 mg
	active		2.5 mg + 14%	1.7−1.9 mg	—
	very active	1.3 mg + 8%	2.9 mg + 32%	2.0−2.2 mg	1.7 mg + 21%
Riboflavine	sedentary	1.6 mg	1.8 mg	2.1−2.2 mg	*1.5 mg − 12%
	moderately active	1.6 mg	1.8 mg	2.2−2.4 mg	1.7 mg
	active		1.8 mg	2.3−2.6 mg	—
	very active	1.6 mg	1.8 mg	2.7−3.0 mg	2.0 mg + 18%
Niacin	sedentary	18 mg		17−18 mg	18 mg
	moderately active	18 mg		18−20 mg	18 mg
	active	18 mg		19−21 mg	—
	very active	18 mg		22−24 mg	18 mg
Ascorbic acid	sedentary	30 mg	75 mg	65−70 mg	60 mg
	moderately active	30 mg	75 mg	70−75 mg	60 mg
	active	30 mg	75 mg	75−80 mg	—
	very active	30 mg	75 mg	85−95 mg	60 mg

*Boxed on USA RDA per 10000 cals and calorie advice in UK.

though this is disputed. The evidence linking vitamin needs directly with calorie intake is good for thiamine but less reliable for riboflavine, niacin and ascorbic acid. The advised intake, depending upon the level of physical activity, is shown in Table 4. If moderate activity is regarded as the standard, then sedentary activity needs are normally about 20% less, the active are advised to take 20% more, and the very active about 40% above those with moderate activity.

Until the last few years it was possible to define the level of activity by direct reference to the person's occupation. Now, however, with the increase in recreational exercise across a broad range of the population (as exemplified by jogging, aerobics, etc.) it is no longer easy to make the distinction purely on the basis of occupation.

Growth

As might be anticipated from the metabolic activity involved, and the anabolic aspect of protein metabolism, vitamin requirements are increased during periods of rapid growth. This is important in children where the needs are substantially increased, particularly when considered on a body weight or food intake basis. The needs are further increased by the tendency of healthy children to be far more active than adults.

Much of the original research relating to vitamin needs was based on observations that a low vitamin intake retards growth. There is now clear evidence from animal experiments and animal husbandry, that higher levels of vitamins substantially increase the rate of growth. The increased stature of young adult humans in most developed countries is now very apparent. How much of this stems from better intake of calories and proteins, how much from higher vitamin intake, is not clear.

Infection and immunity

There is relatively little direct information on the effects of the vitamins in humans on the development of immunity or of the influence of infections on vitamin requirements. However, there is considerable data in animal experiments, and it would appear that much of this has relevance to human nutrition. Among the factors which have been established are that a deficiency of folic acid, pyridoxine and retinol can impair cell mediated immunity, as can a deficiency of ascorbic acid. The first three vitamins all influence T-cell dependent antibody responses but by different mechanisms. Thus pyridoxine deficiency specifically leads to thymic epithelial dysfunction. Retinol has no such effect, but due to the alteration in surface membrane glycoproteins, including those of lymphocytes, retinal deficiency interferes with antigen binding. Ascorbic acid, on the other hand, appears to exert its effect not only via its activity on intercellular material,

13

but also via its effect on glucocorticoid synthesis. There is also evidence that α-tocopherol, via its antioxidant properties, may exert an effect on the development of immunity, particularly when other antioxidants (e.g. ascorbic acid, selenium) are also deficient.

N-nitroso compounds

Considerable attention has been directed recently to the possibility that some N-nitroso compounds may exert a significant carcinogenic role in humans, particularly within the gastrointestinal tract. N-nitroso compounds occur in various food products, in water, alcoholic beverages, cosmetics, industrial products, tobacco products and drugs, among other factors common to normal living. Many of these sources of exposure have now been recognized and action taken to reduce the level of risk. However, in humans endogenous formation of these compounds has been directly demonstrated from various precursors including nitrate, nitrite and other nitrogen containing compounds (Figure 3). Endogenous formation mainly occurs in the stomach, and the extent of the production is increased if the

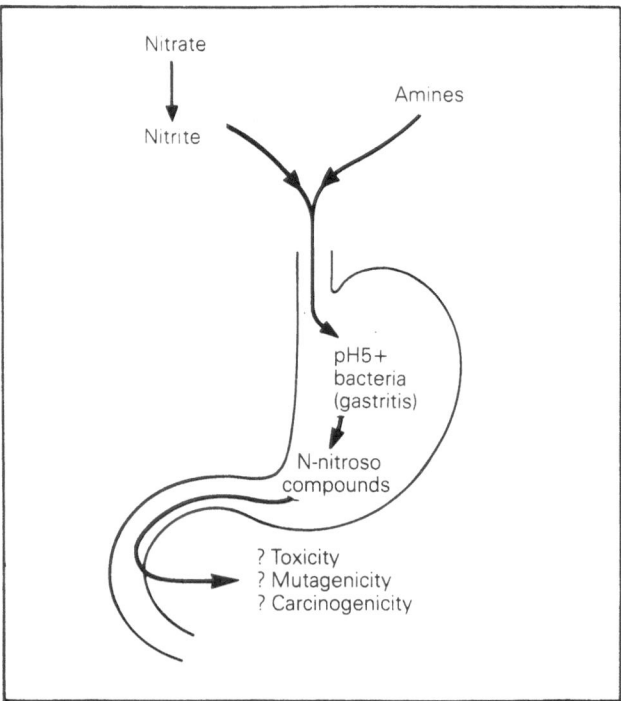

Figure 3 A postulated pathway by which nitrate might be linked to carcinogenesis

level of gastric acidity is low. Hence it is particularly likely to occur in gastritis, in pernicious anaemia and after partial gastrectomy.

It is possible to demonstrate that ascorbic acid and α-tocopherol are effective in blocking intragastric nitrosation, i.e. to reduce the formation of nitroso compounds, though they do not influence the levels of preformed compounds. Since some 10% of adults have a low gastric acidity, this raises the question whether it is logical and desirable to increase the recommended level of intake of these vitamins, particularly ascorbic acid (see *also* under 'Smoking').

Polyunsaturated fat intake

A high intake of fat, particularly polyunsaturated fats, increases α-tocopherol requirements though not in direct proportion to the intake. Indeed there is clear evidence of a tocopherol deficiency clinical state caused by the infusion of polyunsaturated fats. In recent years, as the population has become more aware of the risk factors in the genesis of atherosclerosis the average intake of polyunsaturated fats has risen significantly. Hence it is important that adequate attention be paid to the intake of α-tocopherol.

Smoking

Cigarettes are a significant contributory factor to the level of exogenous nitrosamine (see page 14) in smokers. Non-smokers are also exposed to a similar risk by way of sidestream cigarette smoke. There is also clear evidence that smoking lowers the tissue levels of ascorbic acid. Several reasons exist for this altered vitamin C status, namely: an altered food intake, differences in absorption of vitamin C, the utilization of vitamin C as a detoxifying agent of components derived from smoking and increased metabolic turnover.

Table 5 The effect of cigarette consumption on plasma vitamin C concentration in male subjects

Daily cigarette consumption	Number of subjects	Percentage with plasma vitamin C concentration < 4 mg/l
None	1058	89.5
1– 5	157	91.1
6–10	127	81.9
11–20	399	80.2
21–30	100	71.0
30	33	63.6

Based on Ritzel and Bruppacher (1977)

It has recently been demonstrated that the percentage of subjects with a plasma vitamin C concentration equal to or less than 4 mg per litre varies with the daily cigarette consumption (Table 5). On the basis of the available evidence, it appears that smokers require a daily vitamin C intake which is approximately 40% higher than that of non-smokers in order to maintain a comparable vitamin C status. This is important in the light of an increased exposure to nitrosamines.

Ethnic characteristics

Perhaps the best example of ethnic differences leading to a different vitamin need is that of immigrant Asians in the United Kingdom. A recent survey has suggested that about one Asian in ten has rickets or osteomalacia, while nearly half the Asian adolescents show evidence of subclinical vitamin D deficiency. This probably stems from the limited exposure to sunlight with their pattern of dress; their reduced sunlight exposure in the inclement British weather; and the fact that their largely vegetarian diet does not give them access to some of the normal dietary sources to which the native British are exposed. It is clear, therefore, that when immigration is a significant factor within a community, there is a need to look with care at the recommended levels to determine whether an adequate margin is being provided.

It is interesting to note that the dress adopted by some nations can lead to difficulty even in sunnier climes. A recent report stresses the low vitamin D levels that exist during pregnancy in Saudi Arabia.

Drug interference

Several vitamins are adversely affected, via various mechanisms, as a result of drug activity. This effect, which is considered in detail later (page 105) may significantly alter the vitamin needs.

2

Standard recommendations for vitamin intake

With the different views about how the vitamin needs of the body should be defined, it is scarcely surprising that health authorities in various countries differ considerably in their recommended daily intakes. This variation is also increased because different countries have different views as to the appropriate safety margins. Table 6 summarizes the recommended daily vitamin intake for young male adults undertaking normal levels of activity in six representative countries. The extent of the variation is apparent. Moreover, it will be noted that countries differ as to the vitamins for which they make definite recommendations.

From what has been said above, it follows that there are no currently accepted international standards for recommended levels. Nevertheless, it is necessary to give some general advice. Among the most widely accepted and utilized standards is that produced by the Food and Nutrition Board of the National Research Council in Washington DC, USA. This is updated on a regular basis (normally every 4 years) and published with a good bibliography justifying the conclusions that have been reached. This is, therefore, regarded as perhaps the most logical standard to use at present, and the recommendations for the daily allowances from this United States authority, based on the 1980 figures, are shown in Table 7.

Table 6 Variation in the recommended daily intakes between selected countries (male adults)

Vitamin	USA	USSR	France	FDR	Japan	UK	Range
Retinol (mg)	1.0	1.5	1.0	0.9	0.6	0.75	0.6–1.5
Calciferol (µg)	5	2.5	10	2.5	2.5	10	2.5–10
Tocopherol (mg) (–TE)	10	10	10	12	—	—	10–12
Phylloquinone (mg)	0.07–0.14*	0.2–0.3	—	—	—	—	—
Thiamine (mg)	1.4	1.6–1.7	1.5	1.6	1.0	1.0	1.0–1.7
Riboflavine (mg)	1.6	2.1–2.2	1.8	2.0	1.3	1.6	1.3–2.2
Pyridoxine (mg)	2.2	1.8–2.0	2.2	1.8	—	—	1.8–2.2
Niacin (mg)	18	17–18	18	9–15	16	18	9–18
Folacin (mg)	0.4	0.4	0.4	0.4	—	0.3	0.3–0.4
Cobalamin (µg)	3.0	2.0	3.0	5.0	—	—	2.0–5.0
Pantothenic acid (mg)	4–7*	10	10	8	—	—	4–10
Biotin (mg)	0.1–0.2*	—	0.1–0.3	—	—	—	—
Ascorbic acid (mg)	60	65–70	80	75	50	30	30–80

*Range only quoted

Table 7 Recommended daily allowances of the vitamins according to the Food and Nutritional Board of the National Research Council: Recommended Dietary Allowances, 9th edition. Washington: National Academy of Science, 1980

	Fat soluble vitamins			Water soluble vitamins						
Age (years)	Retinol (µg RE)	Calciferol (µg)	Tocopherol (mg TE)	Ascorbic acid (mg)	Thiamine (mg)	Riboflavine (mg)	Pyridoxine (mg)	Niacin (mg)	Folic acid (µg)	Cobalamin (µg)
Infants										
0–0.5	420	10	3	35	0.3	0.4	0.3	6	30	0.5
0.5–1	400	10	4	35	0.5	0.6	0.6	8	45	1.5
Children										
1–3	400	10	5	45	0.7	0.8	0.9	9	100	2.0
4–6	500	10	6	45	0.9	1.0	1.3	11	200	2.5
7–10	700	10	7	45	1.2	1.4	1.6	16	300	3.0
Adult males										
11–14	1000	10	8	50	1.4	1.6	1.8	18	400	3.0
15–18	1000	10	10	60	1.4	1.7	2.0	18	400	3.0
19–22	1000	7.5	10	60	1.5	1.7	2.2	19	400	3.0
23–50	1000	5	10	60	1.4	1.6	2.2	18	400	3.0
51+	1000	5	10	60	1.2	1.4	2.2	16	400	3.0
Adult females										
11–14	800	10	8	50	1.1	1.3	1.8	15	400	3.0
15–18	800	10	8	60	1.1	1.3	2.0	14	400	3.0
19–22	800	7.5	8	60	1.1	1.3	2.0	14	400	3.0
23–50	800	5	8	60	1.0	1.2	2.0	13	400	3.0
51+	800	5	8	60	1.0	1.2	2.0	13	400	3.0
Pregnancy	+ 200	+ 5	+ 2	+ 20	+ 0.4	+ 0.3	+ 0.6	+ 2	+ 400	+ 1.0
Lactation	+ 400	+ 5	+ 3	+ 40	+ 0.5	+ 0.5	+ 0.5	+ 5	+ 100	+ 1.0

3

The concept of optimum vitamin nutrition

Within the field of animal husbandry it is now widely accepted that improved performance, as determined, e.g. by more rapid growth and healthier animals, results from dietary levels of vitamin intake that are in excess of the minimum required for apparent health (Figure 4). This is particularly noticeable when growth rates are used as the basis for assessment.

In Table 8 the vitamin requirements for various animal species have been extrapolated on a body weight basis to a standard 70 kg man. These levels are compared with those advised for human use in Table 7.

It is fully appreciated that such extrapolation is open to criticism on many counts. Nevertheless, it is interesting to note that with the possible exception of nicotinic acid the recommended levels for man are well below those for other important species when determined on a weight basis.

Table 8 Vitamin requirements for various animal species extrapolated on a body weight basis to a 70 kg man. These are compared with recommended human intakes. Data from the USA has been used throughout

	Vitamin A (i.u.)	Vitamin D (µg)	Thiamin (mg)	Riboflavine (mg)	Niacin (mg)	Pyridoxine (mg)
Man	5000	5	1.4	1.6	18	2.0
Cow (milking)	7500	11	—	—	—	—
Horse (working)	6000	11	3.75	3.0	15	3.75
Pig (sow)	10 500	26	2.25	6.0	22.5	3.75
Dog	25 500	11	6.0	3.0	16.5	1.5
Hen (laying)	12 000	33	5.0	5.0	90	7.4

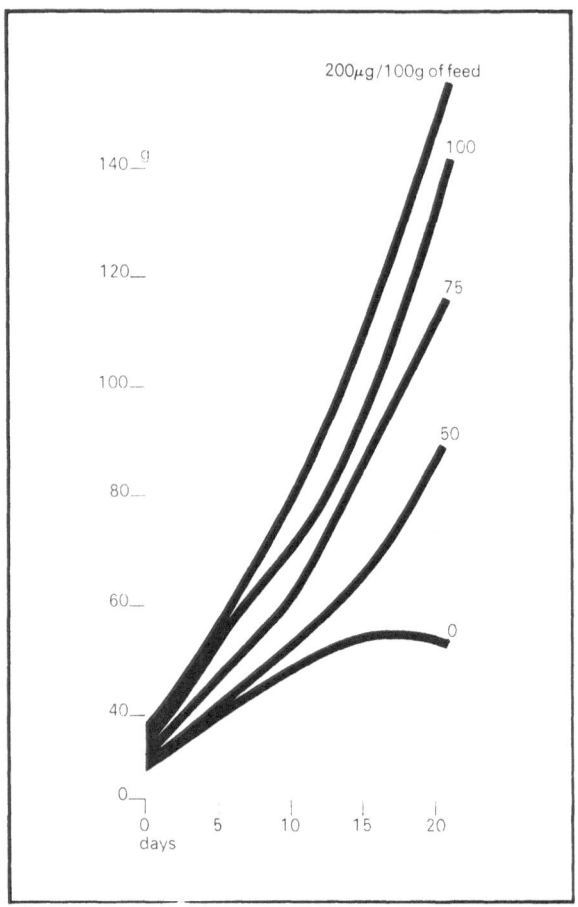

Figure 4 Effect of pyridoxine supplement on growth of chicks

While the concept of optimum dietary intakes is now well established and accepted in animal husbandry, there is currently insufficient data to allow us to establish similar optimum levels for the human. As has been seen above, recent information suggests that there are many factors common to a significant proportion of the population that increase the vitamin needs, and it therefore appears that further investigation of this situation is required. A lack of adequate knowledge of the appropriate intake per day is shown conclusively by the wide range of levels recommended by the authorities in different countries. Hence it would seem to be sensible to consider the desirability of generally higher levels of vitamin supplementation.

4

Overall causes of vitamin deficiencies

The causes of the vitamin deficiencies, whether single or multiple, can be divided into several main classes:

(1) Those associated with an inadequate food intake – normally found in the Third World.
(2) Those associated with an inappropriate intake – as seen in industrially developed countries.
(3) Those associated with inadequate digestion and absorption.
(4) Those associated with increased needs.
(5) Those associated with tissue metabolic disturbances which render a normal intake inappropriate – as seen in certain genetic disorders and in association with other diseases.

These may also be classed as primary or secondary causes. The primary causes (i.e. those involving inadequate intake of the vitamin) embrace categories 1–3 above, while the secondary causes (changed needs) are represented by 4 and 5.

The causes of vitamin deficiency are substantially different between the Third World and industrially developed countries. Hence it is logical to consider them separately (pages 25 and 29).

The causes within industrially developed countries are not experienced in isolation. Certain groups of people (e.g. the elderly, alcoholics) may experience vitamin deficiencies from more than one cause. These are clearly at greater risk than others, and these groups are considered in more detail on page 61 et seq.

An explosive increase in the population coupled with inadequate techniques of food production is leading to a mounting food deficit in many countries, despite the excessive stores of food that exist in others. At

23

present countries fall naturally into two groups in relation to food supply – the 'haves' and 'have-nots', and these equate in general terms to those that are economically developed and those that are not. Whether this remains the relationship, indeed whether the Third World continues to exist in its present form at all, is more a matter for politicians than nutritionists.

Whatever the future may hold, for the present approximately one third of the world's population lives in the economically developed countries with a food abundance. Their average daily intake amounts to about 3000 kcal (about 13 MJ) and the major problem is that of inappropriate eating. On the other hand, the remaining two thirds in the economically underdeveloped countries are subsisting on an average of rather less than 2000 kcal (about 8 MJ) with a substantial number regularly consuming well under this figure.

These two groups differ not only in terms of the quantity of the food but also in its nutritional quality. The greatest difference lies in protein intake and, in particular, the quantity of animal proteins. In the developed countries the daily average intake of protein is almost 40 g per person, while in the underdeveloped countries the level is only about one fifth of this.

Although individual countries may change from one category to another (for example, as a result of the discovery of large fossil fuel reserves) the broad difference between developed and underdeveloped countries is likely to increase over the next few years. The population increase is averaging about 2.1% per annum in the underdeveloped countries, compared with an average of about 1.0% per annum in the developed countries; indeed some are reaching a population plateau or even declining slightly. Moreover, while the developed countries usually have efficient food production and distribution systems they can also afford to import food. On the other hand, food production in Third World countries is often inefficient, yet facilities for improving efficiency or importing the necessary additional food are outside the economic possibilities of the community.

5

Vitamin disorders in the Third World

The *main* nutritional disorders of the industrially underdeveloped countries are those of *protein/energy inadequacy*. Vitamin deficiencies coexist but within the framework of the florid states of starvation are usually, relatively speaking, of secondary importance. Hence the clinical picture that is seen is classically a mixed one, and correction of the protein/energy deficit must be the prime consideration. However, this does not justify not paying adequate attention to the concomitant vitamin deficiency, for as the protein/energy nutritional state improves so the vitamin deficiency assumes a greater importance.

In the underdeveloped countries, three different, though sometimes related, causes explain the poor nutritional state:

Poor climatic conditions

Plates 1–4 show those parts of the world where the greatest nutritional deficiencies occur, and these correspond very closely to those areas where the poor climatic conditions make food production difficult *viz.* the polar regions, the deserts, the arid mountainous areas and to a lesser extent the tropical rain forests. Crop failure is rare in technically developed countries but, on the other hand, failure of the year's crop due to drought, flood or pests is still very common in those areas where climatic conditions are in any case poor and where technical resources are inadequate.

High population density with low economic status

Viewed as a national rather than an international problem, inadequate local food production relative to the size of the community only has relevance if the economic status is such that additional food cannot be imported.

Many of the countries in which there is a relatively high population density are found in the subtropical and tropical regions of Asia. In many of these areas inefficient methods of farming with yields far below the theoretical maximum compound the problem.

Inadequate facilities for health and nutritional education

One of the most difficult problems facing the Third World is that of education in new methods of food production, handling and storage; of the use of new food sources; and of the removal of food taboos. For example, one recent study in Nigeria demonstrated that the standard method of preparing their local vegetables totally destroyed the ascorbic acid which was plentiful in the raw vegetables. The education problem rests not only on an inadequate number of trained personnel, but often on the *laissez faire* attitude of the population which is not helped by their poor nutritional status.

One other difficulty arises from cultural and religious characteristics. Thus for example, even when there is adequate sunlight, as in Saudi Arabia, vitamin D levels are low in both maternal and cord blood (Figure 5).

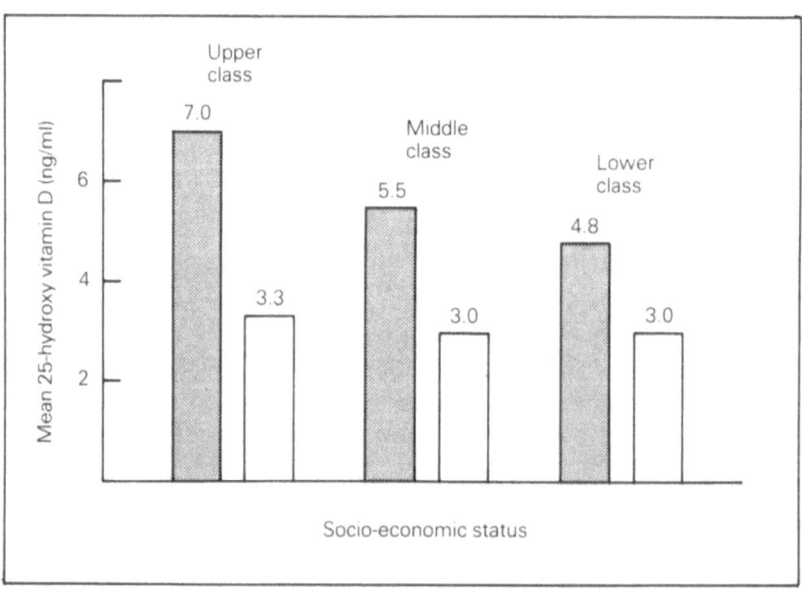

Figure 5 Mean 25-hydroxy vitamin D (ng/ml) in maternal (hatched) and cord (plain) blood from women of different socioeconomic status in Saudi Arabia (based on F. Serenius, A. Alidrissy and P. Dandona *Journal of Clinical Pathology*, **37**, 444–7 (1984))

In consequence of these causes a substantial proportion of the population in the Third World suffers from nutritional deficiencies. The deficiency involves calories and proteins as well as vitamins and minerals. Within the vitamin deficiency area it is common to find multiple deficiencies. Hence it is rare to find the classical signs of deficiency of a single vitamin, e.g. typical patients will often show insufficiency of both vitamin A and several members of the B group. Moreover, nutritional deficiencies lead, in turn, to infections probably by interference with the immune reaction, e.g. measles, tuberculosis, which in turn complicates the clinical picture.

While it is clear that a substantial proportion of the Third World population suffers from nutritional deficiency, there are groups within these communities that are at special risk. These are shown in Table 9. As in industrially developed countries, they are the groups in which the intake is reduced even more than usual, or where the needs are highest.

Table 9 The main groups within a Third World community who suffer from nutritional deficiencies

The elderly
Women
(a) with a large family
(b) during pregnancy
(c) during lactation
Children
Those suffering from infections
Alcoholism

6

Vitamin disorders in industrially developed countries

Although there is an overall adequate diet, there are several causes which can lead to vitamin deficiencies in industrially developed countries.

Inappropriate intake

In certain individuals or groups special factors may lead to diets that are inappropriate in one or more respects. These factors are summarized in Table 10.

Poverty and ignorance

Improvements in social welfare have reduced severe poverty in many countries, but some people are still too poor to buy enough food. Coupled with this poverty there is often gross ignorance of what constitutes a nutritionally adequate diet. Thus the money available for food is frequently frittered away on a diet which leaves much to be desired. Even when earnings increase, the additional money is frequently not spent on food but, in order to satisfy a desire for social prestige, on gaudy clothing, jewellery or bigger weddings. Indeed the diet may even deteriorate in times of greater prosperity, as for example when traditional food is abandoned in favour of convenience foods which are nutritionally less good.

Lack of incentive

People living alone and particularly those who suffer from chronic disease tend to eat convenience foods. While some convenience foods, either naturally or as a result of supplementation, contain adequate quantities of

Table 10 Causes of inadequate vitamin nutrition in developed countries with adequate food production

Due to	Caused by	Population groups mainly affected
Primary food deficiency	poor food storage	hot countries, isolated communities
Diminished food intake	poverty and ignorance	the poor and uneducated, the elderly, alcoholism
	lack of incentive	those living alone
	anorexia	the elderly, anorexia nervosa, takers of anorexic drugs, the sick particularly in chronic disease
	food taboos and fads	religious groups, racial groups displaced from normal environment, pregnancy, fashion conscious dieters, the elderly
	dental problems	the edentulous
Diminished absorption	absorptive disorders	sprue, coeliac disease, chronic intestinal hypermotility etc., stomach and biliary tract abnormality
	parasitic infections	intestinal tract worms
Increased requirements	rapid growth	children, pregnancy
	increased metabolism	labourers, athletes, hyperthyroidism
	infections	chronic infective disease
	drug therapy	specific drugs only, e.g. thyroxin, oral contraceptives, antibiotics
	other dietary components	too high fat/carbohydrate ratio producing ketosis, high polyunsaturated fat diet
	other interfering substances	smoking
Increased losses	diuresis	diuretic drug therapy
	excessive sweating	labourers, athletes
	lactation	nursing mothers

the vitamins, many do not. Thus items such as conserves, bread and cakes figure largely in the diet. This may lead to an excess of carbohydrates and the intake of protein and vitamins may be grossly inadequate.

Anorexia

Anorexia is a frequent precipitating cause of vitamin deficiency in a person whose previous nutritional status has just been adequate. Anorexia is common in the elderly, both in those living alone and in those in an institution. This anorexia results from listlessness, boredom and depression or as a result of feeling unwanted. The loss of wider interests during retirement can cause loss of appetite, and may precipitate malnutrition in a person with a previously borderline nutritional status. This is particularly true for retired spinsters or bachelors.

A further common cause of anorexia is an acute infectious disease. Even a relatively mild infection may cause a decrease in appetite. In breast-fed babies such anorexia not only decreases their immediate food intake, but also by inadequate breast emptying and reduced production of prolactin, leads to a more lasting reduction in the available milk. Thus at the time when nutrition is most important the food supply is least adequate.

Food taboos and fads

Many religious groups have specific and sometimes widespread food taboos. In some cases these taboos stem from sound public health considerations and benefit the community by reducing disease (e.g. avoiding parasite infected meat).

However, many food taboos are without obvious foundation, have a deleterious effect on the general nutritional state of the community, and persist in spite of changes in social structure. Religious taboos not only specify particular items of diet but also define periods of total fast. The occasional day's fast, particularly in an otherwise well-nourished subject, may be beneficial rather than undesirable. However, more extensive fasting, particularly in groups whose nutritional status is in any case suspect, may do harm, e.g. during Ramadan when fasting lasts for one lunar month each year, no food may be taken during the hours of daylight.

Some of the specific food taboos relate to totem observance: some are based on the preservation of the species for sacrificial purposes. Some food taboos have no direct relationship to religious beliefs but to a supposed deleterious effect of a certain food. Many of these food taboos are confined to women, for the majority of the communities were and still are patriarchal societies. The lot of the pregnant woman is especially hard; for not only is she denied all the foodstuffs on the taboo list for women in general, but in

31

addition certain dietary constituents are specifically forbidden during pregnancy in view of supposed adverse effects on the fetus.

It is thus not surprising that such a low intake results in gross nutritional deficiencies at a stage where requirements are highest. It is important to note that vitamin deficiencies in pregnancy are not confined to industrially underdeveloped countries. A recent study in Germany showed that thiamine intake falls well short of the pregnancy RDA while riboflavine, pyridoxine and folic acid are at borderline RDA levels.

The Western World has added fads and fashions in dietary habits to the taboos. Thus, for example there is the low calorie and vitamin intake of the fashion-conscious young and middle-aged woman attempting to slim. In the same age group the fads of the pregnant woman are well known. At the other end of the age scale, many elderly people develop food fads and the majority eat a far from balanced diet.

Dental problems

Dental caries and missing teeth can make eating uncomfortable thus producing a very inadequate and distorted diet. This can occur at any age, but may play a particularly significant role in the elderly.

Apathy

Apathy towards food and little incentive to prepare adequate meals is common in those living alone, the diet may become more and more monotonous and less and less nourishing. Impaired digestion is often associated with the apathy and thus further lowers the nutritional status. This type of problem may be seen in the elderly, in the middle-aged bachelor or spinster living in a one-room apartment with or without adequate cooking facilities and in the teenager living in an apartment for the first time.

Chronic disease

Loss of appetite is frequently associated with chronic diseases. Therefore, at a time when adequate nutrition is especially important (see below) there is an unfortunate mechanism which reduces the nutritional status by decreasing the food intake. Among chronic diseases that commonly lead to inadequate nutrition should be mentioned chronic pulmonary disease, congestive cardiac failure, the effects of a cerebrovascular accident, senile dementia and multiple sclerosis.

Poor digestion and absorption

Absorptive disorders

In the common absorption defect diseases (sprue, idiopathic steatorrhoea, fibrocystic disease of the pancreas etc.) there is inadequate absorption of many dietary constituents. This involves not only the main energy providing constituents, but also the vitamins. In the majority of cases it is the fat soluble vitamins that are least well absorbed, but deficiency of water soluble vitamins (e.g. folic acid) has also been reported.

Parasitic infections

A specific example is seen in the case of infestation by a fish tapeworm which produces a deficiency of cobalamin.

Interactions

Interactions between substances in the intestinal tract and the vitamins may prevent absorption. A typical example is the interaction of the fat soluble vitamins with mineral oil (e.g. liquid paraffin).

The elderly

In the aged, several factors may contribute to impaired vitamin absorption. These include defective mastication of the food, reduction of volume and acidity of gastric secretions, decreased secretion of digestive enzymes in the gastrointestinal tract and changes in the circulation of blood to the digestive tract.

Increased requirements

The causes of increased requirements have already been considered. These causes are summarized in Table 11. If the increased requirements are not met by the diet then a deficiency state will occur.

Table 11 Causes of increased vitamin requirements

Exercise
Growth
Infection and immunity
Other dietary constituents (e.g. N-nitroso production)
Smoking
Drug interactions
Haemodialysis
Ethnic characteristics

Metabolic disturbances

Some tissue metabolic abnormalities can either result in an apparent vitamin depletion or more rarely can produce an apparent excess of individual vitamins.

Apparent depletion can be seen in the following examples: abnormal metabolism of cholecalciferol in certain diseases producing 'resistant rickets' (Table 45); folic acid deficiency resulting from depleted ascorbic acid intake (which reduces the metabolic conversion to folinic acid); increased needs for ascorbic acid among smokers (page 15); pyridoxine deficiency in familial xanthurenic acid disease and cystathioninuria. Examples of apparent excess intake is represented by the extreme sensitivity to calciferol seen in sarcoidosis and nephrocalcinosis (page 131).

7

Clinical manifestations of vitamin deficiencies

Vitamin deficiencies produce clinical effects when they are of sufficient degree to deplete any tissue reserves, leading to a sufficient disturbance of cellular metabolic activity to produce signs and symptoms; at first non-specific (the so-called liminal, marginal or subclinical state – page 77), but subsequently clearly defined and typical. Hence consideration of these clinical effects logically requires a consideration of the activities of the vitamins in the tissues and of the consequent effects of a deficiency state. For some of the vitamins, the relationship between the pathophysiology and the clinical abnormality is obvious; for others the causal relationship is far from clear. For some vitamins apparently minor metabolic abnormalities can give rise to florid clinical effects, while for others, presumably when the metabolic abnormalities are not rate-limiting, severe biochemical disorders may be apparent with only minimal clinical effects. Thus it is not possible to generalize about the vitamins, and each must be considered separately. These individual deficiency states and the clinical effects that are produced are considered in Part 3 (page 111 *et seq*).

8

Assessment of vitamin status

The determination of a vitamin deficiency presents no problem to the clinician when it is present in its classical and overt form, e.g. xerophthalmia. Greater difficulty exists, however, with lesser levels of abnormality (page 77); when there are complicating factors of multiple nutritional abnormalities (e.g. marasmus) a situation which is very frequently encountered, or when other diseases are also present (e.g. infections). A further difficulty arises when the physician fails to realize that a nutritional deficiency may exist in the patient, and fails to search for the appropriate signs. In most industrialized countries, the current but fallacious teaching is that the only nutritional disorder consists of overindulgence in inappropriate foods, and that vitamin deficiencies do not exist. Hence vitamin deficiencies are rarely sought.

The assessment of the vitamin nutritional status of individuals and groups is based upon the use of one or more of the following techniques:

(1) Determination of dietary levels.
(2) Laboratory estimation of the tissue state.
(3) Clinical examination.

When a pronounced vitamin deficiency produces the overt and classical clinical signs and there are no complicating other disorders to confuse the picture, the diagnosis will be obvious. However, this is rare in clinical practice. In any case the physician should try to reach the diagnosis before the stage of florid clinical manifestations is reached.

In borderline cases, any of these techniques used in isolation may give misleading or inadequate information and the best, though expensive, results are achieved by the simultaneous use of at least two independent methods of assessment.

Determination of dietary levels

The main method that has been used to determine the dietary intake is that termed the 'shopping basket' (in some countries, 'market basket') survey, and despite the limitations this is still the most widely used technique. If this is applied to the individual patient, he (or she) is asked to record the food that they purchase over a period of time, both in terms of description and quantity. The method can also be used for a defined population group by the determination of the overall food intake and then averaging it over the group. The intake of nutrients based upon the 'shopping basket' survey system is usually extrapolated for content on the basis of standard tables. However, this is rarely accurate. The main reasons for the inaccuracy are shown in Table 12.

Table 12   The main causes of discrepancy between nutritional intake calculations based on 'shopping basket' surveys calculated by tables of food contents when compared with food weight measurement and chemical analysis of an aliquot

Fluctuation in the content of the fresh natural raw food

Losses during storage:
- (a)   natural unprocessed stored food,
   e.g. rodent, insect pest, bacterial and fungal contamination
- (b)   foods processed prior to storage,
   e.g. losses in blanching, leaching and sterilization

Presence of nutrients in a non-absorbable form

Presence of inedible portions of the food

Losses during food preparation before and during cooking

Plate waste

In addition to the factors shown in Table 12, the surveys which are concerned with an average of the overall food usage (i.e. the total quantity of the food consumed divided by the total population) take no account of the great variability of the eating habits of individuals or groups. Even the amount of food entering the kitchen does not give a representation of that consumed, for it makes no allowance for wastage in preparation or from the table. It must be appreciated that not only are these losses variable but that such variation is cumulative so that the overall inaccuracy of such surveys can be substantial (Table 13).

A second and more direct way of determining what an individual eats is by direct questioning. This removes many of the problems related to variability of eating habits encountered in the 'shopping basket' survey, but

Table 13 Representation of the cumulative inaccuracy that may occur in the 'shopping basket' survey calculation of ascorbic acid intake from potatoes

Example 1	Total ascorbic acid (mg)	Example 2	Total ascorbic acid (mg)
Main crop freshly dug	300	main crop stored 8–9 months	80
Wastage due to deterioration (nil)	300	wastage due to deterioration (10%)	72
Peeling (thin) − 5%	285	peeling (thick) − 10%	65
Boiled and served directly (− 30%)	200	boiled and kept warm (− 60%)	26
Final intake	200	final intake	26

Hence a portion of 'boiled potatoes' in a 'shopping basket' survey may contain between about 26 and 200 mg ascorbic acid

still suffers from many difficulties and disadvantages (Table 14), particularly if skilled and experienced questioners are not used for the surveys.

The third related technique is the one which gives the greatest accuracy, but requires competent participants. It involves the weighing of the individual ingredients of the meal on an adjustable balance on which the actual meal plate can be placed. Even with this technique there is the difficulty of accounting for the food which is left on the plate at the end of the meal. The technique is very tedious for the patient and expensive to undertake due to the time taken on the interpretation of the content of prepared food.

But even if a reasonably accurate record can be taken of the reputed intake, difficulties exist when these are converted into chemical values. These difficulties include the inherent inaccuracies of dietary content tables which are referred to later. They also include the variability of inedible

Table 14 Some problems encountered in dietary history surveys

Subject says what he believes he ought to eat and not his actual diet

Main meals are described, but snacks, drinks and sweets are ignored

There are accidental or intentional lapses of memory

Quantities are badly assessed

'Made-up dishes', e.g. fruit cakes, may have greater variability of composition than is usually recognized

The same name refers to different foods in different areas, and conversely the same foods are given different names, making recognition difficult

proportions from one food to another (e.g. the average wastage of fruit and vegetables is in the range 20–60% of the amount purchased, of dairy products 10% or less). But the extent of the wastage depends not only on the nature of the food but the predelections of the person studied. This latter effect is compounded by preparation and plate wastage which depends on the affluence of the society; the national and local food customs and on the skill of the cook. Examination of the swill bucket will show how marked this loss can be, particularly with institutional type meals.

However, in spite of the inherent difficulties and inaccuracies, calculation based upon the estimated intake and tables of composition is still the most widely used method of assessing the dietary intake, and gives results that are adequate for most investigation and nutritional control purposes.

When greater accuracy is required it is necessary to weigh each constituent of the diet as it is put on the plate, and then, if necessary, to subtract the weight of any uneaten remains, a much more time-consuming procedure which requires that the trained person be present at all meal times and when any snacks are taken. An educated person can often undertake the food weighing themselves, but if this is done it is essential to ensure that there are no reasons why deliberately dishonest results might be anticipated (e.g. in certain patients with obesity).

These weighed portions are usually translated into chemical content by the use of appropriate food composition tables, which exist now for most areas of the world and which commonly give figures for both individual foods prepared and cooked in different ways and for a selection of compounded foods. Several nutritional units have also prepared computer programs so reducing the tedium of the calculation. However, it is important to appreciate that cooking methods change the nutritive value of the food. This change represents the complicated interaction of the effects of, for example, heat treatment reducing the water content, and the vitamin reduction that results both from the high temperature and the leaching out of the water soluble fractions in the cooking water. Hence care must be taken that the correct food entry is examined.

Nutritionists can be broadly divided, into those that swear by the food tables and those that swear at them! Their true value lies somewhere between these two extremes, for they give a reasonable guide to the nutritional value of the diet. If accurate measurements are required for the intake of the vitamins (or indeed any dietary constituent), the only method is to weigh each food accurately just before it is eaten, and to perform direct chemical analysis on an aliquot of each item. This is a laborious and expensive procedure. The extent of the difference between analysed and calculated intake is shown in Figure 6. If losses in processing and lack of availability of the natural form for absorption are also considered, recent work suggests that errors in some of the published figures amount to an overstatement by factors of up to 10:1.

40

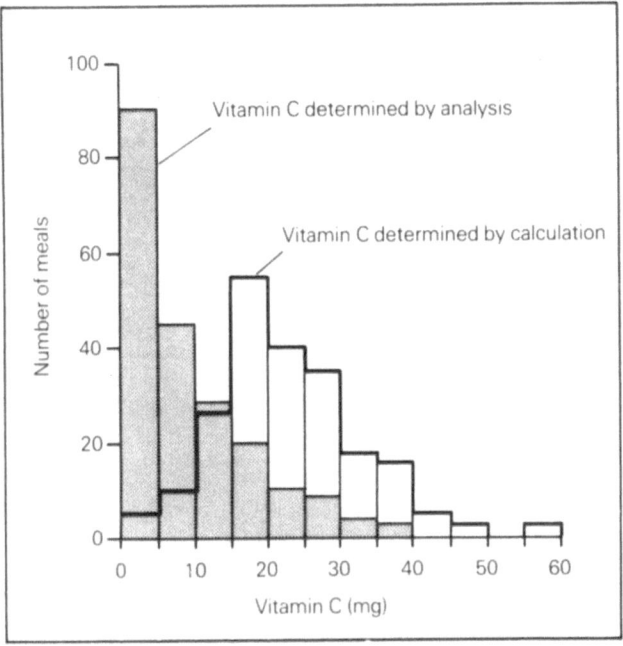

Figure 6 Comparison of the calculated (plain) and analysed (hatched) values of vitamin C in 200 meals-on-wheels. (Based on *Three Score Years . . . and Then*. L. Davies. William Heinemann Medical Books, London, 1982)

The original vitamin assays of foods, before their chemical nature was known, was usually based upon the determination of the specific effects of the foods in deficient animals, and the tests could be divided into curative and protective types. These techniques determined true availability, but were expensive and the data were not necessarily relevant to another species. The specific tests that now exist for most of the vitamins can be divided into four classes: chemical, microbiological, animal protective, animal curative. Detailed accounts of the exact assay methods are beyond the scope of this book and are available in the standard analytical reference books. Table 15 summarizes the current main methods for each of the vitamins. Several of these methods require prior preparation of the vitamin from the food by chemical, chromatographic or other methods, and the extent of losses in this process should also be considered.

Laboratory estimation of the tissue state

There are three distinct methods for determining the tissue vitamin status.

Table 15 Summary of the main assay methods for each of the vitamins in food, divided according to chemical, microbiological, animal or receptor assays

Vitamin	Chemical	Microbiological	Animal assay	Receptor assay
Retinol	colorimetric, fluorimetric	—	yes	—
Calciferol	ultraviolet (UV) absorption*, colorimetric, fluorimetric*, gas liquid chromatographic (GLC)	yes	yes	—
Tocopherol	colorimetric, UV absorption, oxidimetric, fluorimetric, GLC	—	yes**	—
Phylloquinone	colorimetric*, fluorimetric*, oxidimetric polarographic, GLC	—	yes	—
Thiamine	fluorimetric, polarographic**, UV absorption	yes**	yes**	—
Riboflavine	photometric*, fluorimetric, polarographic	yes	yes	—
Pyridoxine	colorimetric*, polarographic*, fluorimetric*, GLC	yes	yes**	—
Niacin	colorimetric, fluorimetric**, polarographic, GLC	yes**	yes**	—
Folic acid	photometric*, fluorimetric*	yes	—	—
Cobalamin	photometric, fluorimetric, radiometric, polarographic, atom absorption spectroscopic	yes	—	—
Pantothenic acid	colorimetric, GLC	yes	yes**	—
Biotin	photometric, GLC	yes	yes**	—
Ascorbic acid	colorimetric, oxidimetric, polarographic, GLC	—	yes	—

* Can only be applied to food assay with great difficulty and loss of accuracy
** Accurate but only rarely used

(1) *Degree of body saturation by:*
     (a)   blood or plasma levels
     (b)   urinary excretion.
     However, the vitamin in the plasma is not filling a metabolic function. It is being transported to the tissues where it will be metabolically active. Hence the plasma status or urinary excretion is not necessarily a reliable index of the tissue status.

(2) *Tissue level studies*
     Estimation of tissue desaturation by direct analysis is the only method which gives a true indication of the tissue vitamin status, but this measurement of tissue levels is often difficult and even this level may not represent the active form.

(3) *Metabolic studies*
     Within the tissues, vitamins exert their effects by influencing metabolic activity. Sometimes this will be as the vitamin itself (e.g. vitamin C), sometimes in converted form (e.g. vitamin D), but often after conversion to a coenzyme (e.g. vitamin B group). Tissue vitamin levels only have relevance when equated with the metabolic requirements. The tests for biochemical metabolic efficiency attempt to determine the vitamin level of the tissues not in isolation but relative to their metabolic needs. Some tests measure the active metabolic form of the vitamin, others determine the metabolic consequences of the vitamin level. This will sometimes be based on a natural process that occurs within the cell and sometimes on the influence of the vitamin level on an administered test substance.

A summary of these laboratory methods is considered in more detail on page 89 in Part 2.

Clinical examination

Specific symptoms and signs exist for most of the vitamin deficiencies though, as is explained elsewhere (page 81), multiple deficiencies or complicating factors commonly obscure the exact clinical picture. The question of clinical disorders has already been considered (page 35), and the classical signs of the deficiency states of the individual vitamins are described in Part 3. From this description it is apparent that certain areas of the body will provide the greatest amount of information on possible vitamin deficiencies. Attention should be directed particularly to these when poor nutrition is suspected.

The whole question of clinical examination is further considered, and a decision tree for the management of suspected vitamin deficiencies is included in Part 2 (page 85).

9

Therapeutic uses of the vitamins

The administration of vitamins for therapeutic purposes can be considered under three main headings.

(1) Prophylaxis for suspected or proven inadequate vitamin intake and use in preventive medicine.
(2) Therapy when an inadequate vitamin intake leads to clear clinical evidence of a deficiency.
(3) Pharmacological actions of the vitamins in doses which are in excess of those required for the relief of deficiency states.

Prophylaxis

Vitamins should not be regarded just as substances administered to correct deficiencies for it is more logical to try to prevent the disorders. This can be done either by general supplementation of the diet or by specific administration of one or more of the vitamins to those who fall into the risk groups.

Supplementation of the diet is most logical when there is a general low level of intake such that a large proportion of the population is at potential risk. It may either be done for the whole population or for a specific segment. This can be achieved by choosing the type of food that is to be vitamin supplemented. In such a food selection it is important to consider not only the foods that are most acceptable to the target group, but also whether any interaction occurs between the vitamin and the food constituents which would prevent adequate levels of the vitamin being absorbed. Based on the considerable information now available it is possible to select both the appropriate food and the level of supplementation.

Administration to the individual is more appropriate when there is a small but clearly defined risk group. In these circumstances general administration to a selected or whole population is wasteful.

Apart from the question of administration of vitamins to avoid deficiency states, recent information suggests that some of the vitamins may themselves have other protective effects. Among these may be mentioned the use of vitamin E for the prevention of deterioration in the storage of fats for human consumption; the possible use of higher levels of vitamin C in the prevention of the formation of nitrosamines under risk circumstances (smokers, ingestion of foods containing nitrates (page 15), and the postulated role that vitamin A and its analogues might play in the prevention of cancers. These uses are still mainly experimental but initial studies are encouraging.

Therapy

The therapeutic uses are considered for each of the individual vitamins in turn in Part 3, and mixtures of the vitamins may be administered for particular purposes (e.g. for alcoholism). A therapeutic index is given in Part 2 (page 95 et seq.). Moreover substances which are chemically related to the vitamins may also be used for therapy (e.g. the retinoids, page 120; cholecalciferol related substances, page 131; folate derivatives, page 174).

One recent specific clinical nutritional development is total parenteral nutrition, mainly for severe gastrointestinal disorders. The institution of such therapy should be left to the specialized gastroenterologist who should ideally work with a nutritionist. However, these patients are now being discharged from hospital on such therapy, and it therefore behoves practitioners to understand at least the possible problems that may be involved. These fall into two categories.

First, when vitamins are absorbed from the intestinal tract there are mechanisms that reduce the extent of the absorption. No such mechanism exists when the vitamins are administered directly into the parenteral circulation. The other protective mechanism that is bypassed by parenteral administration is the metabolic function of the liver. Hence it is possible to achieve vitamin overdosage with parenteral vitamin nutrition.

Second, our knowledge of vitamin requirements is still far from perfect, particularly in relation to other diseases or drug treatments. Hence vitamin deficiencies are periodically reported when currently accepted daily levels are infused. In consequence the recommended levels for infusion are modified from time to time, and further modification may be expected in the future. The currently advised levels (based mainly on the views of the American Medical Association) are shown in Table 16.

Pharmacological actions

The therapeutic value of the vitamins in deficiency states is abundantly clear. More controversial are their possible effects as pharmacological

Table 16 Composition of multivitamin preparation for use with total parenteral nutrition (based on American Medical Association advice for those over 11 years old). Expressed as daily intake

Retinol (i.u.)	3300
Calciferol (i.u.)	200
Tocopherol (i.u.)	10
Thiamine (mg)	3.0
Riboflavine (mg)	3.6
Pyridoxine (mg)	4.0
Niacin (mg)	40
Pantothenic acid (mg)	15
Folacin (μg)	400
Cobalamin (μg)	5.0
Biotin (μg)	60
Ascorbic acid (mg)	100

agents in the absence of evidence of deficiency. Such use only applies to certain of the vitamins and these are considered in Part 3 and in the section on the therapeutic index (page 95).

Routes of administration and dosage

The route of administration and the dosage level which should be used will depend upon many factors, including the weight and age of the patient, their clinical state, the urgency that exists for correction of the deficiency, the absorption capacity of the intestinal tract, and certainly not least the availability of appropriate pharmaceutical forms. This availability differs markedly from one country to another and even from one year to another in the same country. For all these reasons it is not feasible to define either the route of administration or the dosage in absolute terms.

It is, however, useful to provide an indication of the factors that lead to the selection of one route of administration or another, and to indicate the appropriate range of dosage that applies to the majority of patients seen in clinical practice. This advice should be read in association with the information given in the therapeutic index (page 95 *et seq.*) and in Part 3 for the individual vitamins. In these latter areas deviations from the normal dosage levels for specific disorders are noted.

Selection of route of administration

The following principles apply:

(1)   Most vitamins are well absorbed from the intestinal tract and oral therapy is adequate. The exceptions to good absorption are:

In malabsorption syndromes.
In the presence of vomiting or diarrhoea.
For vitamin B_{12} in pernicious anaemia.
Presence of substances in the diet that interfere with absorption.

(2)   Parenteral administration should normally be used only if:
There are problems with absorption.
A rapid action is required.

(3)   Most commercially prepared parenteral preparations can be used by either the intravenous or deep intramuscular route. Intravenous injections are best avoided if possible, but if they are used it is desirable to administer the dose very slowly. Allergic reactions may occur with vitamin B_1 intravenous injections and with some formulations of fat soluble vitamins (due to the solubilizer).

(4)   In those countries where administration by suppository is one of the preferred routes, it can be accepted that rectal absorption is good for most of the water soluble vitamins but may be irregular for the fat soluble ones.

Dosage

Table 17 summarizes the dosage range for the regular use of vitamins, depending upon whether a prophylactic or therapeutic use is required, it defines a low and high dose level and gives values for infants and children where these appear to be applicable. It should be noted that the dosages given do not cover the high level needed for some pharmacological actions.

Indications for multivitamin therapy

One of the principles of modern therapy is to administer specific single substances according to the therapeutic need. A mixed preparation is only regarded as appropriate when there is a clear indication of synergy between the constituents. Vitamin therapy should, in general, follow the same principle, and in consequence it is important to define those disorders for which a multivitamin preparation is appropriate, i.e.:

The elderly

From what has been said in various parts of this book it may be seen that there is a high incidence of vitamin inadequacy in the elderly (e.g. page 65; Tables 26–28). For some the deficit is such that a frank clinical deficiency will be present; for others there will be a marginal deficiency (page 77) which requires therapy.

Table 17 Common routes of administration and advised dosages. Unless otherwise stated daily dosage is quoted

Vitamin	Route	Prophylaxis	Low dose	Therapy High dose	Infants	Children	Units
Vitamin A	oral	5000	5000–30 000	100 000–200 000	5000	up to 15 000	i.u.
	i.m.	300 000 6 monthly	50 000 monthly	150 000 monthly	—	up to 50 000 monthly	i.u.
Vitamin D	oral	500–1000	5000	10 000–15 000	up to 5000	up to 10 000	i.u.
Vitamin E	oral	3–15	10–100	400–600	5–25	up to 100	mg
Vitamin K$_1$	oral	5–10	5–10	10–20	5	up to 10	mg
	i.v./i.m.			100 slowly			mg
Vitamin B$_1$	oral	10–20	5–10	100–600	1.0/kg	up to 5	mg
	i.m./i.v.	3–10	10–100	100 slowly	up to 5	up to 20	mg
Vitamin B$_2$	oral	1–5	5–10	10–100		up to 10	mg
Vitamin B$_6$	oral	20	20–50	50–300	up to 3	up to 100	mg
Niacin	oral	10	10–100	100–500	5	up to 20	mg
Vitamin B$_{12}$	i.m.	25–50 monthly	100–250 weekly	up to 500 weekly		—	μg
Folic acid	oral	2.5–10	5–10	10–20		5–15	mg
Pantothenate	oral	100	100–200	200–600		up to 100	mg
	i.m.			500–1000			mg
Biotin	oral	—	10		2–5	5–10	mg
Vitamin C	oral	100	100–200	up to 3000	30–50	up to 500	mg

In the elderly it is unusual for the deficiency to be confined to only one vitamin, indeed most show a deficiency of vitamin C, several members of the vitamin B group and vitamin D. In view of this widespread vitamin inadequacy, a multivitamin preparation is the logical form for therapy or prophylaxis.

The institutionalized

Those who are institutionalized for long periods usually experience a deficiency of several vitamins. Hence the use of a multivitamin preparation is logical.

Children

Attention has been drawn already (page 13) to the fact that growth increases the requirements for many vitamins. It is widely held that the average mixed diet supplies adequate levels of the vitamins to cope with such extra needs. However, many recent studies (page 66; Table 30) have demonstrated that levels of intake in children and teenagers are often not adequate. It is probable that this results, at least in part, from the fact that such children do not eat a balanced diet, but consume a large quantity of readily available and palatable pre-packed snack foods.

When there is doubt about the vitamin adequacy of the diet in this group, a multivitamin preparation is desirable as several vitamins are usually depleted.

Pregnancy

During pregnancy there is not only an increased metabolic load on the mother, but a substantial additional requirement to allow for the rapid growth of the fetus. This is recognized in the higher levels of intake of many vitamins recommended by most national authorities for pregnancy. Typical supplements for pregnancy above those advised for normal women are shown in Table 18. There are, however, some specific vitamin needs in pregnancy.

It has been suggested that folic acid deficiency may be a cause of a circumvallate placenta which predisposes to abruptio placentae. Adequate folate intake appears to help to prevent this complication, which has certainly become much less common since folate supplements have been used more extensively. Neural tube defects are also found more commonly in women in the lower social class, and mothers who give birth to such infants have been found to have lower body levels of folic acid and other vitamins. Clinical studies have suggested that vitamin supplements that include folate reduce the incidence. For these purposes, the folate intake should be at least 10 mg daily.

Table 18 Recommended vitamin supplements in pregnancy and lactation (i.e. to be added to the normal adult female requirement)

	Pregnancy		Lactation	
	Amount	% Adult female need	Amount	% Adult female need
Retinol (RE)	0.2 mg	25	0.4 mg	50
Cholecalciferol	5 µg	50	5 µg	50
Tocopherol	2 mg	25	2 mg	25
Thiamine	0.4 mg	40	0.5 mg	50
Riboflavine	0.3 mg	25	0.5 mg	40
Pyridoxine	0.6 mg	30	0.5 mg	25
Niacin	2 mg	15	5 mg	40
Folic acid	400 µg	100	100 mg	25
Cobalamin	1.0 µg	33	1.0 µg	33
Ascorbic acid	20 mg	33	40 mg	66

Since there is a need for supplementation of many vitamins, it is logical to administer the supplement in the form of a multivitamin preparation. Special formulae providing the pregnancy supplements including folate are available in many countries.

Lactation

As with pregnancy, so with lactation. There are substantial additional nutritional requirements to compensate for both the increased metabolic activity and for the vitamin content of the milk. Supplements for lactation are advised by most national authorities and those recommended by the USA authorities are shown in Table 18. Since the requirement is for supplementation with several vitamins the use of a multivitamin preparation is logical.

Alcoholism

Alcoholism is now one of the most common causes of vitamin deficiency in industrially developed communities (page 71). The deficiency seen is multiple, involving not only thiamine, but also all the fat soluble vitamins, folic acid, cobalamin, ascorbic acid and pyridoxine. A recent report has indicated that though these are the vitamins for which a deficiency can be established, the levels of other vitamins may be borderline. Thus for example when alcoholism was treated with high doses of thiamine and pyridoxine, pellagra (niacin deficiency) was precipitated.

Hence it is vital to treat all patients with chronic alcoholism with high doses of a multivitamin preparation to ensure that the broad range of deficiencies will be treated.

Malabsorption syndromes

It is usual for the malabsorption syndromes to affect not just one vitamin but to involve the fat soluble vitamins and some of those that are water soluble (particularly folic acid and vitamin B_{12}). Hence an appropriate multivitamin preparation is indicated for supplementation, although it is important to ensure that the preparation that is used has adequate levels of the fat soluble members of the group.

Drug-induced deficiencies

The interactions between drugs and vitamins are considered on page 105. From this it is apparent that while some drugs interfere with single vitamins, usually at the receptor or enzyme site, there are others, particularly those that influence absorption from the intestinal tract that tend to produce a deficiency of several members of the series. When this latter situation occurs it is logical to use a multivitamin preparation rather than administer the individual substances.

Cancer patients

For many years interest has focused on a possible relationship between low levels of certain of the vitamins and the development of neoplasia. There is clear evidence that retinol, ascorbic acid and tocopherol are involved with various immune functions in the body and that cell mediated immunity is one factor in the growth or otherwise of neoplastic cells. The relationship of ascorbic acid to the endogenous production of nitrosamines (which may have a role in the genesis of neoplasia) has also been established (page 14). Retinol intake and tissue ascorbic acid levels have also been shown to be lower in those who have developed lung cancer.

There is no evidence that vitamin supplementation directly affects the course of the disease, but since there are needs for additional tissue repair and for an effective immunological defence it does seem to strengthen the case for, at least, supplementation of those vitamins that are concerned with these effects in such patients. Maintenance of nutritional integrity can certainly do no harm and may produce some benefit. Since there is a requirement for several vitamins it is logical to use a multivitamin preparation.

Recovery after severe illnesses

There is clear evidence that many severe illnesses can deplete the vitamin reserves. This is particularly true in infections treated by broad-spectrum antibiotics. In such patients, because there is usually a deficiency of several vitamins, a multivitamin preparation can be used with benefit.

Anorexia

Lack of appetite, whether it be due to physical illness, social causes or anorexia nervosa, results in undernutrition for several vitamins and is best dealt with by the administration of a multivitamin preparation.

Haemodialysis

This therapeutic procedure leads to a general reduction of the water soluble vitamins (but elevation of those that are fat soluble). A water soluble multivitamin preparation (containing the vitamin B group plus vitamin C) is convenient.

Supplementation in low birth weight infants

In another part of the book (page 73), it has been demonstrated that there are extra, but currently not fully established, needs for vitamins among premature babies, particularly those of very low birth weight. Supplementation is essential and the currently advised levels are shown in Table 19. Despite the fact that most of the vitamins must be added, individual vitamins should be used at the appropriate level and not a standard multivitamin preparation.

Table 19 Suggested allowances of vitamins that are most needed in low birth weight babies

Vitamin D_3	800–1000 i.u. daily orally or possibly calcitriol 0.02–0.03 µg/day i.m.
Phylloquinone	1–1.5 mg at birth repeated if necessary at 10 day intervals
Ascorbic acid	50 mg daily
Folic acid	100 µg daily
Pyridoxine	100 µg daily
Vitamin E	3 mg daily

10

Vitamin safety

Since, by definition, the vitamins are essential for the life and well-being of animals and man their ingestion in quantities equivalent to those found in a normal mixed diet must be beneficial. Hence at a level of intake adequate for their vitamin role adverse reactions are unknown except in a very few clinical disorders.

However, in recent years the vitamins have been administered in increasingly high doses, often with dramatic therapeutic benefit. In consequence, cases (often anecdotal) of alleged adverse reactions have been reported.

It is, therefore, necessary to define the safety of the vitamins. We are concerned not with normal intake, but with levels far in excess of these which are sometimes administered to obtain the pharmacological actions.

A designation of vitamin safety that covers all possible circumstances is very difficult. Firstly, there are the natural differences from one individual to another. Secondly, this takes no account of the additional variations imposed by idiosyncrasy, hypersensitivity or illness. Hence the spread of vitamin levels which may lead to adverse reactions is large. Thirdly, the period of ingestion and the reason for the intake of the vitamins differs. At one end of the spectrum lies the daily consumption of vitamin enriched food; at the other end the physician monitored therapeutic prescription of high doses for a limited period.

Bearing these difficulties in mind, the most rational approach to defining safety is to compare the highest level at which adverse reactions do not appear with the Recommended Dietary Allowances (RDA) as utilized in the USA. This has been taken as an internationally widely accepted standard (Table 7, page 19). For ease of comparison the values for males age 23–50 are summarized in Table 20.

The level of safety of the individual vitamins is considered in the relevant portion of Part 3. However, it is possible to draw some general conclusions.

Table 20 The Recommended Dietary Allowances (RDA) based on NRC Food and Drug Board advice 1980. The figures for males (aged 23–50) are shown for comparison with the figure for safety. For further details see Table 7

	RDA (males aged 23–50)
Thiamine	1.4 mg
Riboflavine	1.6 mg
Pyridoxine	2.2 mg
Niacin	18 mg
*Pantothenic acid	4–7 mg
Folic acid	400 µg
*Biotin	100–200 µg
Vitamin B_{12}	3 µg
Ascorbic acid	60 mg
Vitamin A	1000 µg
Vitamin D	5 µg
Vitamin E	10 mg
*Vitamin K	70–140 µg

* Estimate of 'safe and adequate' level rather than RDA due to the limited evidence available

* The levels of vitamins normally ingested by the majority of the general population in the diet do not exceed the Recommended Dietary Allowance by a factor of more than two and are safe.

* The risk of adverse reactions appears to be greatest when high doses are taken without professional advice.

* An acute intoxication is very rare and encountered only in: the idiosyncratic reaction to parenteral thiamine (page 148); the neurotoxic reaction to massive doses of pyridoxine (page 158); flushing with nicotinic acid (page 162); adverse reactions to vitamin A (page 119) and vitamin D (page 131).

* There is a considerable margin of safety with most of the vitamins, for there are very few side effects attributable to the vitamins at dose levels even substantially above the Recommended Dietary Allowance (Table 21).

* With the sole exceptions of adverse reactions from ingested vitamins A (page 119) and D (page 131), even the rare cases of vitamin side effects that occur are rapidly reversible on dose reduction and leave minimal or usually no lasting effects.

Table 21 Representation of the ratio between the established high levels that can still be regarded as safe in the vast majority of adults and the RDA for normal adults. Exact figures are not possible, but the code utilized represents the following approximate ratios: + up to 10 times, + + 10–50 times, + + + 50–100 times, + + + + at least 100 times the RDA. See text for the evidence upon which these estimates are based

	Ratio 'high safe'/RDA
Thiamine	+ + +
Riboflavine	+ + + +
Pyridoxine	+ + +
Niacin	+ + +
Pantothenic acid	+ + + +
Folic acid	+ + + +
Biotin	+ + + +
Vitamin B_{12}	+ + + +
Ascorbic acid	+ + + +
Vitamin A	+ +
Vitamin D	+
Vitamin E	+ + + +
Vitamin K	+ +

PART 2

PRACTICAL ASPECTS

11

Risk groups and evidence of dietary vitamin inadequacy

From what has been said in the general description of the causes of vitamin deficiency (pages 23–34) it can be seen that there are two distinct geographical areas in which nutritional abnormalities can occur: the Third World and the industrially developed world: and that the aetiology of the nutritional disorder differs significantly from one to the other.

Within the Third World the majority of the population is potentially at risk of a vitamin deficiency. Hence it follows that physicians working in these communities should always be on the lookout for the symptoms and signs directly attributable to vitamin deficiency. This aspect of Third World deficiency has been covered in the earlier part of the book (page 25) and needs no further attention here.

In the industrially developed world, however, there is a widely held, but fallacious, view that vitamin deficiencies are rare. In consequence physicians tend to ignore the possibility that symptoms and signs may be related to vitamin deficiencies and miss their significance.

There have been several recent surveys which have investigated the vitamin status of the general population in various countries (mainly industrially developed). Some of these surveys are based on a determination of the proportion of the population receiving an intake less than the RDA, and some on the proportion of the population who show biochemical or clinical evidence of deficiency. A summary of these surveys is given in Table 22. The vitamin intake inadequacy is apparent. Reference can also be made to some surveys which show deficiencies of specific vitamins (e.g. vitamin A page 114, Table 42; vitamin C page 187, Table 55).

Two extensive studies have been undertaken recently in the USA. Tables 23 and 24 show that a substantial proportion of the general population of that country have an inadequate vitamin intake.

61

Table 22 Surveys indicating inadequate vitamin intakes in the general adult population (see also Tables 42 and 55 for surveys of vitamins A and C)

Country	Type of study	Result	Reference
United Kingdom	Dietary determination	Folic acid average 213 µg – (53% RDA); Range 168–254 µg – (40–60% RDA)	Poh Tan et al., 1984
Thailand	Population survey	Thiamine 25% of population are significantly deficient	Vimokesant, 1979
Canada	Population survey	Ascorbic acid 9% of eskimo population show signs of scurvy	Mongeau, 1980
France	Hospital inpatients	Vitamin B$_1$ Below RDA in 57% Vitamin B$_2$ Below RDA in 47% Vitamin B$_6$ below RDA in 53% Vitamin C below RDA in 9%	Lemoine et al., 1980
Germany	100 industrial workers	Vitamins B$_1$, B$_2$, B$_6$ folic acid 52% are deficient in at least one of these vitamins	Hotzel, 1979
Finland	1175 15–36 year old normal people	Vitamins B$_1$, B$_2$, niacin and A 90% showed intake below RDA	Harju, 1979
Greece	Population survey	Niacin – in Athens 56 new cases of pellagra were seen over the last decade	Stratigos and Katsambas, 1982
Mexico	Population nutrition survey (2248 persons)	Niacin 82% were below RDA Vitamin A 98% were below RDA Vitamin D 94% were below RDA	Chavez et al. (in press)

Table 23 Distribution of food intake among 37 785 US normal individuals – based on Nationwide Food Consumption tables

Vitamin	Percentage between 70–100% RDA	Percentage less than <70% RDA
Vitamin A equivalent	19	31
Vitamin B_1	28	17
Vitamin B_2	22	12
Preformed niacin	24	9
Vitamin B_6	29	51
Vitamin B_{12}	19	15
Vitamin C	15	26

Table 24 Percentage of the normal US population with intake of less than 70% RDA of selected nutrients. Based on the Second National Health and Nutrition Examination Survey (NHANES II)

Vitamin	Male	Female
Vitamin A	41.8	44.5
Vitamin C	30.0	36.5
Vitamin B_1	22.1	30.5
Vitamin B_2	13.5	23.9
Niacin	14.0	22.1

However, in addition to vitamin inadequacy that can be established in a proportion of the general population, certain groups of people are at greater risk of developing multiple or single vitamin deficiencies (Table 25).

Table 25 The main groups at risk of vitamin deficiencies in industrially developed countries

With extensive evidence of deficiency:
     The elderly
     Children and teenagers
     In pregnancy and during lactation
     Alcoholics
     Those in institutions
     Smokers

With a priori evidence of deficiency:
     Dieters
     Those living in 'bedsitters'
     Low birth weight infants
     Patients of haemodialysis
     Those receiving some medications

Table 26 Vitamin nutritional surveys: risk groups – elderly patients

Population group (number)	Country	Vitamins	Comment	Ref.
Elderly patients (80)	UK	B group, C	90% showed biochemical deficiency, most clinical signs as well	Exton-Smith & Scott, 1968
Elderly patients	UK	C	23% showed low leukocyte levels and positive Hess test	Eddy, 1971
Elderly patients (81)	UK	Folic acid	12% showed reduced red cell folate	Varadi & Elwis, 1966
Elderly subjects (72)	UK	Folic acid	40% showed low serum folate	Hurdle & Williams, 1966
Elderly subjects (451)	UK	C	50% had intake below RDA	Milne et al., 1971
Elderly in rehabilitation	South Africa	B_1, niacin, B_6, B_2, C	$\quad$ % <RDA $\quad$ Bioch def. B_1 $\quad$ 94% $\quad$ 38% niacin $\quad$ 87% B_6 $\quad$ 100% $\quad$ 66% B_2 $\quad\quad\quad$ 28% C $\quad\quad\quad$ 62%	Labadarios & Rossouw, 1981
Elderly	New Zealand	C, B_1	60% deficient in C and B_1	McLean et al., 1976
Elderly patients	Switzerland	A, B_1, B_6, C	majority had intake less than 50% RDA	Schlettwein-Gsell, 1975
Elderly (124)	US	B_1, B_2, B_6	B_1 biochem def – 6.5% B_2 biochem def – 13.5% B_6 biochem def – 2.0%	Barnes, 1983
Elderly (7)	US	B_1	B_1 biochem def – 14.2%	Bayonini & Rosalki, 1976
Elderly (234)	US	B_1, B_2	B_1 biochem def – 6.0% B_2 urinary excretion – 0%	Brin et al., 1965
Elderly (153)	Germany	B_1, B_2, B_6	B_1 biochem def – 22.9% B_2 biochem def – 11.7% B_6 biochem def – 19%	Hoorn et al., 1975
Elderly on admission to geriatric units (1082)	UK	D	3.7% showed biopsy evidence of osteomalacia	Campbell et al., 1984

The elderly

There have been many studies of the vitamin nutritional status in the elderly utilizing nutritional, clinical and/or biochemical criteria (Table 26). They almost all show a significant proportion with evidence of vitamin deficiency.

In one study in the United Kingdom, the Department of Health reported on a multicentre trial in the elderly. They studied the food intake and looked for biochemical evidence of deficiency (and borderline nutrition) for several of the vitamins among other constituents of the diet (Table 27). They also

Table 27 Survey of the vitamin intake and status in the elderly in the UK (based on DHSS 1979). Food and biochemical evidence of poor intake

Vitamin	Food survey percentage below RDA			Biochemical evidence of low* levels (%)	
	Male	Female		Male	Female
A	54	54		Not tested	
B_1	87	85		8	
B_2	52	58		30	
Niacin	84	81		Not tested	
B_6	98	100	serum	12.2	7.5
			red cell	3.4	5.5
C	84	89	plasma	14	
			leuk	18	
D	89	89		Not tested	
Folic acid	not tested		serum	13.3	13.5
			red cell	4.7	6.0
B_{12}	not tested		serum	2.4	2.6

* Low levels are based on true abnormality. The quoted figure ignores the borderline proportion (about 20%) in each case

Table 28 Survey of the elderly in the UK. Clinical abnormalities normally attributed to vitamin deficiency (based on DHSS 1979)

	% Whole survey (n = 365)	% Group defined as malnourished (n = 26)
Angular stomatitis	2.2	0
Cheilaris	2.5	3.8
Smooth atrophic tongue	4.4	0
Sublingual haemorrhage	6.0	30.8
Red/seborrhaeic nasolabial area	4.9	15.3
Pigmentation exposed skin	20.3	46.2
Hyperkeratosis	9.6	34.6
Sublingual varices – Micro	35.6	69.2
Gross	7.7	23.1

clinically examined the total group of patients and a subgroup defined as clearly malnourished, and reported *inter alia* on the proportion who showed clinical signs normally attributed to vitamin deficiency (Table 28).

The majority were consuming less than the RDA (and often substantially below it) for most of the vitamins. Between 3 and 30% showed biochemical evidence of deficiency of various vitamins, and another approximately 20% showed a marginal biochemical state. A substantial proportion showed clinical abnormalities.

The vitamins widely accepted as being deficient in the elderly include vitamin B_1, niacin, folic acid, vitamin B_{12} and vitamin D. Vitamin C and vitamin B_{12} levels correlate negatively with the cognitive status in the elderly, indicating that low tissue levels of these substances (and thiamine and folate) may represent significant factors in the genesis of geriatric cerebral deterioration.

The main factors which are responsible are shown in Table 29. It is usual for several factors to act together to produce a nutritional deficit in an individual. Sometimes these factors are inter-related: thus limited mobility may make the elderly housebound which can lead to loneliness, social isolation and depression. This in its turn leads to apathy, anorexia and poor food intake. Not only does this poor intake lead to a vitamin deficiency of a level which can lead to tissue depletion and metabolic abnormalities, but there follows a significant incidence of clinical disorders directly attributable to the vitamin deficiency which in their turn perpetuate the problem.

Table 29 Primary and secondary causes of malnutrition in old age

Primary	Secondary
Poverty	impaired appetite
Ignorance	dental problems
Social isolation	malabsorption problems
Physical or mental disability	increased requirements
	alcoholism or interfering drugs

Children and teenagers

Another group that is specifically at risk are children and teenagers (Table 30). Probably the main cause here is rapid growth, but poor dietary habits (snack meals) probably play a role.

Pregnancy and lactation

The additional load imposed by the growth of maternal and fetal tissues and

66

Table 30 Vitamin nutrition surveys: risk groups – children and teenagers

Population group (number)	Country	Vitamins	Comment	Ref.
5–16 year old children of Asian immigrants	United Kingdom	D	11% of children with rickets	Goel, 1979
Ten year olds	Nigeria	A, B$_2$	Average daily intake as % RDA Vit A 75% RDA Vit B$_2$ 80% RDA	Nnanyugo, 1981
Pre-school children	USA	A, C	Abnormal laboratory tests 1–2% vit A 5–25% vit C	Owen et al., 1974
Children	USA	A, C	2–20% abnormal vit A 2–5% abnormal vit C	Garn & Clark, 1975
Children under 15	Columbia	Folate, B$_{12}$	7.2% low serum folate 2.7% low vitamin B$_{12}$	Solon et al., 1978
Young adults	France	B$_6$	In 100% subjects intake below RDA	Brubacher et al., 1978
Teenagers	Austria	B$_1$	9% deficient; 45% marginal	Brubacher, 1979
18–20 year old recruits	Switzerland	B$_1$, folate	B$_1$ – marginal in 36% Folic acid – marginal in 31%	Stransky, 1980
Secondary school teenagers (19)	Nigeria	B$_2$	84% showed biochemical evidence of B$_2$ deficiency	Ajayi, 1981

67

Table 31 Vitamin nutrition surveys: risk groups – pregnant women

Country	Vitamins	Comment	Ref.
France (11)	B_1, B_6, D	In 100% of the subjects the intake was below RDA	Brubacher et al., 1978
UK (240)	Folic acid	Megaloblastic marrow changes in 25% of 'healthy' pregnant women	Chanarin & Rothman, 1971
UK (195)	B_1, B_2, C, D, A	Percentage below RDA B_2 – 62% C – 54% D – 100% A – 46%	Smithells, 1980
Switzerland	B_1, B_{12}, B_6, C, A, folic acid, biotin, B_2	Percentage deficient B_1 – 36% B_{12} – 32% B_6 – 29% C – 27% folic acid – 24% biotin – 16% A – 14% B_2 – 5%	Dostalova (in press)
Germany	B_1	25–30% deficient	Heller, 1974

68

the subsequent vitamin losses in the milk are sufficient to ensure that the mother who does not receive vitamin supplements is very likely to show depletion of at least one vitamin and usually several (Table 31). The ones most likely to be involved are folic acid, riboflavine and pyridoxine.

This depletion is now recognized in most industrially developed countries, and the administration of multivitamin supplements during pregnancy and lactation is now the rule rather than the exception.

Dieters

At the present time it is usual for up to 20% of the population of industrially developed countries to be dieting. The proportion is even higher for young and middle aged women.

The method of losing weight varies widely from total fasting through published menus to self-encouraging groups employing restriction of the intake of normal food.

Whether or not the normal diet provides adequate levels of all the vitamins, it is clear that food restriction without vitamin supplementation is certain to produce a significant reduction in intake. The extent of the reduction has recently been the subject of an American study which looked at the vitamin content of ten of the most popular reducing diets. The results are shown in Table 32. None of the diets met the USA RDA for all the vitamins studied, and six of the ten showed reduced levels in several vitamins. The vitamins that were most likely to be deficient were B_1, B_6 and B_{12}.

Hence unless the diet used is one based on recommended levels (e.g. The Cambridge Diet), vitamin supplements should be advised for all dieters.

Those living in institutions

Institutional catering has a reputation in many countries of predictable awfulness, whether it be in schools or general or long-stay hospitals. When, as in many long-stay hospitals, there are the added factors of economic stringency, a community who cannot supplement the diet elsewhere and the effect of drugs (page 105), malnutrition is all too common.

There is one recent report from the United States in which it was found that half of all the hospital patients studied were suffering from some degree of malnutrition, and that between 5% and 10% died due to starvation.

While the malnutrition which excites the greatest attention is that involving calories and protein, there are numerous studies that have shown both a depleted vitamin content of the diet and evidence of a vitamin deficiency. The vitamin deficiencies of clinical significance most likely to be found are those of vitamin B_1, folic acid, vitamin C and vitamin D.

The problem may be compounded when the institutionalized patient is

Table 32 Vitamin content of ten popular reducing diets

		Average daily calories	Vitamin A	Vitamin C	Thiamine	Riboflavin	Niacin	Vitamin B_6	Folacin	Vitamin B_{12}
Cambridge	Low	330	100	100	100	100	100	100	100	100
Pritikin 700		737	325	786	69	98	88	79	157	35
Simmons		924	70	165	64	83	105	50	56	51
Beverly Hills		928	461	798	56	50	40	96	405	0
I Love New York		980	386	551	60	125	126	73	106	291
Scarsdale	Calories	1014	222	555	57	71	115	65	84	37
F-Diet		1241	291	317	153	171	175	140	283	43
Pritikin 1200		1273	511	928	110	136	118	114	196	42
I Love America		1307	197	403	91	106	110	60	90	56
Stillman		1316	52	34	36	92	181	64	32	764
Atkins	High	2031	92	92	70	94	107	67	64	110

Based in part on data from La Chance and Fisher (1983), Rutgers University.

suffering from epilepsy. For some time it has been known that several anti-epileptic drugs can lead to folate deficiency (page 172). It has now been shown that both folate and biotin levels in epileptic patients show a negative correlation with the total amount and average dose of anticonvulsants. There also appears to be a higher risk of riboflavine and pyridoxine deficiency in such patients, though neither of these correlate with the anticonvulsant. The pyridoxine reduction may be particularly relevant in view of its importance in the synthesis of γ-amino-butyric acid, a natural brain anticonvulsant (page 157).

Alcoholism

Of all the patients who present in the practitioners office in industrially developed nations, among the most likely to demonstrate manifestations of vitamin deficiencies are those with an alcohol problem.

Most people who are moderate drinkers but eat an adequate diet are not at exceptional risk for nutritional deficiencies. Alcoholics, who often consume as much as 20% of their total calories in the form of alcohol place themselves at risk for various vitamin deficiencies. The extent of the risk is demonstrated by the fact that lesions indicative of Wernicke's encephalopathy were found in no less than 1 in 8 of recognized alcoholics in one recent study. The most common manifestations (Figure 7) were in the brain

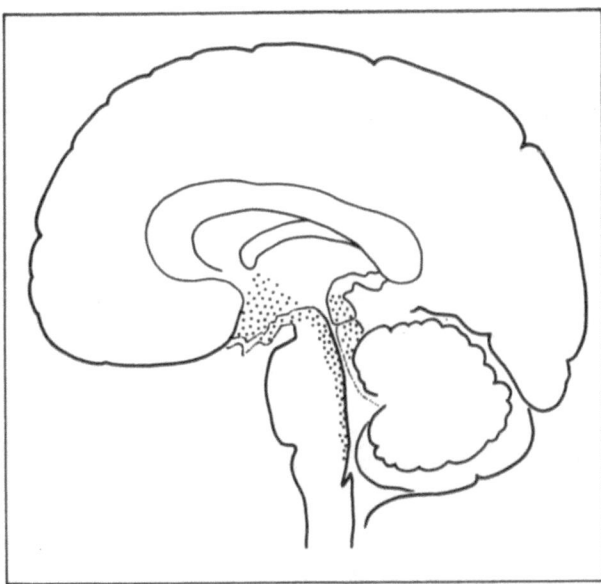

Figure 7 Cross-section of the brain showing the areas affected in Wernicke's encephalopathy

stem, and cerebellar atrophy was also seen. In one third of those who showed autopsy evidence of this thiamine deficiency, it was clear that there had been an active disease at the time of death; yet it had been diagnosed in only 3% of the whole group, i.e. only 10% of the 'active' group. The majority of the remainder of the patients who demonstrated autopsy evidence of inactive Wernicke's disease showed dementia or were permanently institutionalized at the time of death. Other recent studies have confirmed this substantial incidence of alcoholic nutritional brain damage, undiagnosed during life.

The cause of the disease is not just a substitution of the normal diet with a source of calories with no vitamin content, but a vicious circle involving poor appetite, inadequate food, inappropriate food demanding high vitamin levels, gastrointestinal disturbances that reduce absorption, and liver damage that reduces the storage and conversion of the vitamins to their active forms (Figure 8).

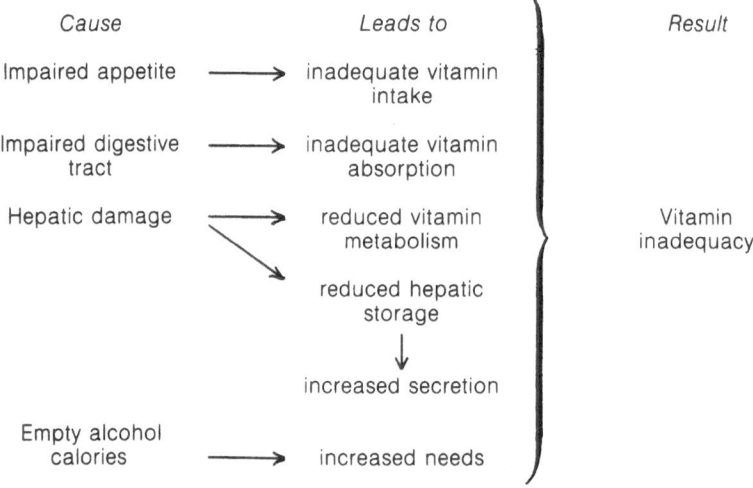

Figure 8 The factors leading to metabolic abnormalities in alcoholic patients

Although thiamine deficiency is the one that is best recognized in alcoholics, deficiency of many other vitamins has also been reported. The list of vitamin deficiencies at a level that causes concern for health includes:

thiamine; all the fat soluble vitamins (retinol, calciferol, tocopherol and phytomenadione); folic acid; pyridoxine; ascorbic acid; cobalamin.

The main clinical manifestations in alcoholics that have been ascribed to the vitamin deficiencies are shown in Table 33.

Table 33 Signs of vitamin nutritional deficiency encountered in chronic alcoholics

Face	nasolabial seborrhoea	vitamin B_2
Lips	angular stomatitis	vitamin B_2
Tongue	red raw tongue with atrophic papillae	niacin
Gums	haemorrhages	vitamin C
Eyes	conjunctival xerosis	vitamin A
	capillary suffusion	vitamin B_2
Skin	perifollicular hyperkeratosis	vitamin A
	xerosis and brittle nails	vitamin A
	curled hairs	vitamin C
	petechiae and haemorrhages	vitamin C

The 'bedsitters'

It is not only the elderly who live alone and therefore suffer from the difficulty which arises from self-catering. A substantial proportion of the younger population of industrially developed countries live alone in apartments or 'bedsitting rooms'. A significant proportion of this accommodation is equipped with only the minimum of cooking facilities. In consequence of this and the lack of incentive to prepare meals for one person, convenience foods are the order of the day.

The fact that this is not just a theoretical possibility is shown by two recent studies. One of these examined the vitamin intake of self-catering students who were slimming. They found that the intake of pyridoxine, folic acid and vitamin D were substantially below the RDA. But this aspect was not due just to slimming, for the other study which investigated the pyridoxine levels of healthy US university students found that all showed biochemical evidence of deficiency.

Low birth weight babies

For some time it has been appreciated that premature babies may have special nutritional problems and needs, tocopherol being a notable example. As the medical management of the premature improves, and hence the mortality rate falls and smaller infants can be saved, so the importance of nutritional aspects assumes greater importance. The factors which predispose to different needs include:

(1) *Decreased body stores.* The placenta is an inefficient mechanism for the transmission of many of the vitamins. Fetal stores appear, therefore, to be laid down late in fetal life.

73

(2) *Increased general utilization*. This is necessary for the increase in growth, for bone calcification, for the high protein and poly-unsaturated fat intake in the special diets that are provided and for red cell development.

(3) *Decreased intake*. This may apply whether the infant is being fed an artificial diet orally or is receiving intravenous nutrition. When feeding is by the oral route there is the additional complication of poor intestinal absorption in the very premature.

(4) *Specific disorders of the premature*. Perhaps the most important of these that is currently recognized is the respiratory distress syndrome – the main cause of mortality in the low birth weight infant. Since this leads to poor tissue oxygenation, oxygen is often supplemented and this may lead to retinal damage in infants with inadequate tocopherol levels. Another recently discovered problem is that related to the poor hydroxylation capacity of the premature, calciferol is only poorly converted to calcitriol and some specialist units are advising that calcitriol should be administered rather than calciferol. Such a use should currently be restricted to these specialist units where blood levels can be monitored.

To date the exact needs of low birth weight babies are not clearly known, and further information and modification of the nutritional advice expected as experience increases. At present the most critical needs in the prematures appear to be calciferol, tocopherol and phytomenadione among the fat soluble, and ascorbic acid, folic acid and pyridoxine among the water soluble. Current advice on suggested levels of administration are given in Table 19 (page 53).

Smoking

Some 50% of the adult population smokes and smoking depletes the tissue levels of vitamin C (page 15). Hence smokers are one of the risk groups, specifically for vitamin C.

Haemodialysis

Another important, though small, group who demonstrate vitamin problems are those receiving treatment by haemodialysis. This procedure leads to reduced levels of the water soluble vitamins but elevated levels of those that are fat soluble. A supplement of water soluble vitamins is therefore required.

Those receiving other medication

Some medications interfere with the vitamin status by a variety of mechanisms. These problems are considered in detail on page 105 *et seq.*

Overall risk level

In most industrially developed countries the safety margin of vitamin levels in the average diet will allow for one of the above adverse factors without evidence of reduced reserves. However, when two or more of the factors (page 24) act together, the reduction may be such that clinical deficiency may be seen. We cannot assume that living in a developed country with an overall average good food intake implies that there is no risk of vitamin deficiency to certain groups of people.

12

Marginal deficiencies

Vitamin deficiency which leads to clearly defined and recognized clinical abnormalities represents the end stage of a slide through depleted reserves and metabolic abnormality to tissues changes that become apparent. There is now little dispute about the existence and significance of such overt clinical disorders nor of the need for appropriate therapy.

Equally there is now clear evidence about the stages through which the vitamin economy declines when the balance between needs and intake is negative. The main dispute which still exists, and which is likely to continue for some time, is the clinical significance of the stage of depletion before classical signs of deficiency occur and specifically whether:

This can lead to complaints of ill health.
This subclinical state requires therapeutic attention.

The first two stages, prior to the development of the frank, clearly distinguished clinical syndromes is now commonly referred to as the stage of *marginal, liminal or subclinical deficiency*. At this stage we can determine the specific biochemical and metabolic abnormalities, and it is during the stage of metabolic abnormalities that the first *non-specific* clinical symptoms of malaise and illness occur. The symptoms are largely behavioural in nature. They consist of, for example, loss of weight, loss of appetite, general malaise, increased irritability, somnolence.

These symptoms cannot incontrovertibly lead to a diagnosis of vitamin deficiency because they are so non-specific. Nevertheless, there is evidence that they can be directly caused by tissue metabolic disturbance due to vitamin deficiency at a level at which overt physical signs are not produced. The evidence for this is:

(1) Those who have worked extensively in the nutritional field recognize these symptoms as being present in clinically diagnosable vitamin deficiencies.

(2) When vitamin deficiencies have been produced under controlled laboratory conditions in volunteers (e.g. vitamins C and B_1, pantothenic acid), these non-specific symptoms appear at the stage when metabolic disturbances can be found with no clinical signs.

(3) At the stage of metabolic disturbance, there is clear evidence of alteration of drug metabolism, which must be regarded as having clinical significance.

(4) Vitamin C depletion to the metabolic disturbance stage leads to the following manifestations:
    (a) Experimental wounds heal more slowly;
    (b) In young adolescents, the working capacity, as determined by oxygen utilization, was proportional to the plasma vitamin C level (Table 34) and the administration of vitamin C improved the working capacity until a plasma vitamin C level of about 0.8–0.9 mg/dl was reached (Figure 36, page 175). This plasma vitamin C level corresponds to a daily intake of about 80–100 mg.
    (c) Correction of marginal ascorbic acid levels can lead to an increase in the haemoglobin level.

(5) If a patient with vitamin A deficiency at the marginal stage is given an increased amount of food with no change in the amount of vitamin A it can precipitate xerosis.

(6) Marginal vitamin deficiencies can lead to poor immunological responses under experimental conditions.

(7) Marginal deficiency of folate can lead to hypersegmentation of the leukocytes before there is any evidence of an anaemia.

Table 34 The oxygen intake (l/min) during exercise in subjects divided according to the level of vitamin C in the plasma. (Based upon Buzina et al., Cartegena symposium 1984)

Vitamin C level (mg/dl)	00–0.19	0.20–0.59	0.60–0.99	>1
Number	32	50	26	63
Oxygen intake (l/min)	2.81 + 0.22	2.86 + 0.42	3.13 + 0.38	3.14 + 0.31

Based on Buzina et al. (1984)

It must, therefore, be concluded that the stage of tissue metabolic abnormality does represent a significant disorder and that the symptoms which occur at this stage should be regarded seriously.

The question then arises whether there is any need for therapeutic action in the marginal deficiency stage. Most nutritionists now agree that appropriate treatment should be instituted at this stage because:

(1)   In the elderly, particularly those in institutions, a significant number of marginal deficiency cases are seen during any survey. Treatment of the group surveyed leads to improvement in the non-specific symptoms in those with the clinical signs.

(2)   Since any stress is likely to precipitate a frank clinical disorder in those with a liminal deficiency it is good medical practice to avoid a disease by treatment.

(3)   Iatrogenic disease due to drug administration is recognized as a current significant medical problem. Since it is clear (page 105) that vitamin levels significantly alter drug effects the maintenance of a normal vitamin status is highly desirable.

On the basis of all these considerations we may, therefore, conclude that the presence of a marginal vitamin deficiency should be regarded as medically important and in need of adequate and appropriate therapy.

13

Clinical diagnosis of vitamin deficiencies

In the Third World, particularly during the many periods of severe food restriction, the question of a nutritional deficiency affecting vitamins as well as calories and protein is always in the doctor's mind and clinical diagnosis should be clear.

On the other hand, in industrially developed countries there is a fallacious view that nutritional disorders are rare. In consequence they may not be sought and hence will not be found, maintaining the fallacy of non-existence.

Clinical manifestations

As with any clinical disorder, there are no short cuts to the clinical diagnosis of vitamin deficiencies. Accurate diagnosis demands a clear history (Table 35) coupled with a full clinical examination. There can be few doctors who have not at some time fallen into the trap of jumping to a diagnostic conclusion on inadequate evidence and subsequently regretted their mistake.

It is, on the other hand, clear that there are certain signs that are indicative of vitamin deficiencies, and the clinical examination should seek for these when the history suggests that there may be a vitamin imbalance. These signs are summarized in Table 36. A decision that no clinical evidence of a vitamin disorder exists cannot be made until at least all these signs have been excluded after a positive search.

Table 35 Symptoms that *may suggest* vitamin disorders and that should lead to careful clinical and/or laboratory examination

Site	Symptoms	Possible vitamin disorder
General	fatigue, malaise, apathy, depression	many vitamin deficiencies (particularly B group) hypervitaminosis A
	anorexia	vitamin B_1 deficiency hypervitaminosis D
Nervous	headache	hypervitaminosis A
	tingling, numbness, burning skin	vitamin B_1 or B_2 deficiency
	low back pain	folic acid, vitamin B_{12} deficiency
	ataxia	vitamin B_1 or B_{12} deficiency
	personality changes	niacin deficiency
Eyes	poor night vision, eyes feel dry	vitamin A deficiency
	blurred vision	vitamin B_1 deficiency
Mouth	bleeding gums	vitamin C deficiency
	lips hurt	vitamin B_1 deficiency
	tongue sore, poor taste	vitamin B_2, niacin, folic acid deficiency
Skin	ecchymoses	vitamin C deficiency
	dry skin	vitamin A deficiency
Gastrointestinal	diarrhoea	niacin deficiency
Cardiovascular	dyspnoea	vitamin B_1 deficiency
Bones	pain	hypervitaminosis A

NOTE: There are of course many other causes for these symptoms but vitamin disorders should be considered

Decision tree for clinical management

The diagnosis of any disorder which leads on to the administration of appropriate therapy follows a natural path. For most doctors, experience allows the steps in decision making to be completed rapidly and accurately. This is particularly true if the patient is well known to the doctor, and who is therefore aware of the background history. However, it is useful to have a decision path available, at least as an *aide memoire* to the existence of vitamin deficiency disorders, or for the puzzling case. A structured approach to the diagnosis of vitamin deficiencies leading to appropriate management is useful and follows:

Is there reason to believe that the food intake is low or inappropriate?

Yes – see below

No – see page 84

Table 36 The clinical manifestations of vitamin deficiency

Site	Sign	Possible vitamin deficiency
Skin	dry skin	? vitamin A
	yellow coloration	vitamin B_{12}
	petechiae, ecchymoses	vitamin C or vitamin K
	follicular hyperkeratosis	vitamin A
	spiral and unerupted hairs	vitamin C
	scrotal/vulval dermatitis	vitamin B_2 or niacin
	pellagrous dermatitis	niacin
	burning feet	pantothenic acid
	pallor	folic acid or vitamin B_{12}
Face	nasolabial seborrhoea	vitamin B_2 or vitamin B_6
	malar pigmentation	niacin
Lips	angular stomatitis	vitamin B_2
	cheilosis	vitamin B_2 or vitamin B_6
Tongue	magenta coloured fissured	vitamin B_2
	beefy red swollen	niacin
	soreness, glossitis	vitamin B_{12}, folic acid, or vitamin B_6
Gums	spongy bleeding	vitamin C
	gingivitis	niacin
Eyes	Bitot's spots, xerosis	vitamin A
	keratomalacia	vitamin A
	poor dark adaptation	vitamin A
	corneal vascularization	vitamin B_2
	amblyopia	vitamin B_1 or niacin
	intraocular haemorrhage	vitamin C or K
	optic neuritis	vitamin B_1 or vitamin B_{12}
Intestinal tract	diarrhoea	vitamin B_1 or niacin
Skeletal system	craniotabes	vitamin D
	frontal and parietal bossing	vitamin D
	epiphyseal swelling	vitamin D
	rickety rosary	vitamin D
	painful subperiosteal haematoma	vitamin C
Nervous system	Wernicke's dementia	vitamin B_1, vitamin B_{12}, niacin, or folate
	Korsakoff's psychosis	? vitamin B_1
	muscle wasting and weakness	vitamin B_1 or B_6
	sensory loss	vitamin B_1 or B_6
	foot drop	vitamin B_1
	reduced tendon jerks	vitamin B_1
	subacute combined degen.	vitamin B_{12}
Cardiovascular system	heart enlargement	vitamin B_1
Blood	anaemia, hypochromic	vitamin C or K
	anaemia, normochromic	vitamin B_2 or B_6
	anaemia, macrocytic	vitamin B_{12} or folic acid

The main causes of an overall poor intake of food or of inappropriate food within the industrially developed countries is given on pages 29–35 and the main groups who are at risk from this point of view are given on page 61 et seq.

Action

If there is reason to believe that there may be dietary inadequacy then it is important to establish whether this has actually led to a vitamin deficiency or whether it has at least reduced the vitamin stores. The assessment of the vitamin status by laboratory tests and clinical examination are considered in general on page 81 et seq. and under the individual vitamins.

In general, however, where there is an inadequate or inappropriate diet, a multiple vitamin deficiency is likely, involving both the fat and water soluble components. Ideally an accurate diagnosis of the vitamins that are depleted should be undertaken. However, this is rarely necessary in practical therapeutics because the vitamins are substances with a wide margin of safety, and for which the use is relatively inexpensive compared with the biochemical determinations that would be required. Hence where there is presumptive evidence of inadequate or inappropriate food intake, management should realistically be by the administration of adequate doses of a multivitamin preparation.

Is there evidence of interference with the vitamins in the intestinal tract?

Yes – see below

No – see page 85

There are relatively few causes for disturbances of vitamin nutrition that can be ascribed to interference within the intestinal tract. The few that exist are considered on page 33. This possible cause of vitamin disturbance can normally be considered very rapidly.

Action

Such causes will normally lead to a deficiency of an individual vitamin, and this should be confirmed before therapy is contemplated. In such patients it is better to treat the cause rather than supplement the vitamins for the long term. Hence the cause should if possible be removed and a short course of the appropriate vitamin provided to restore the vitamin economy to normal.

Is there evidence of an intestinal absorption defect?

Yes – see below

No – see page 85

The most common disturbance which can lead to poor absorption from the intestinal tract is any severe and prolonged diarrhoea. If this is due to an infection, then it is the water soluble vitamin status that is likely to be affected first since there are relatively poor stores of these vitamins. If, on the other hand, the diarrhoea is due to any of the malabsorption disorders it is likely that the fat soluble vitamins will be most affected, although some of the water soluble ones will also be involved.

Action

In any case of diarrhoea it is essential to determine the cause and then try to remove it. This is true for both infections and for malabsorption disorders. Most are now capable of being cured.

If the case is an infection and a cure is affected rapidly, then a short course of a multivitamin preparation, particularly one which supplies the water soluble vitamins may be all that is necessary. On the other hand, if there is a malabsorption disorder, then supplementation with the fat soluble vitamins and with selected members of the water soluble vitamins and certain minerals may be necessary either in the short term or on a long term basis. If long term therapy appears to be required then the author believes that it is desirable to determine the needs accurately and to ensure adequate levels of the selected vitamins. On the other hand, the author accepts that accurate determination is time-consuming and costly and accepts the view held by many, that supplementation by a multivitamin preparation with adequate levels of at least the fat soluble vitamins and folic acid may be more cost-effective.

Is there evidence of increased vitamin need?

Yes – see below

No – see page 86

The main causes of increased vitamin needs are considered on page 11 *et seq.* For the majority, the diet will be adequate to meet these needs and vitamin insufficiency will be unlikely. Where, however, there are reasons to believe that the intake is borderline, then increased needs may lead to problems.

Action

When there is an apparent increased need it is important to assess whether the need is likely to involve the majority of the vitamins (e.g. growth, pregnancy) or whether the effect will be confined to only one or a few vitamins (e.g. pyridoxine dependency syndrome).

If it is likely that the increased needs apply to many vitamins then there are good reasons for supplementation with a multivitamin preparation at adequate dosage without defining exactly which of the vitamins are involved. On the other hand, when it appears that there is a specific need it is important to check the vitamin status carefully before deciding on the appropriate therapy. The methods that are available for testing the vitamin reserves and the tissue metabolic activity related to the vitamins are considered on page 89.

Is there clinical evidence of a vitamin deficiency?

Yes – see below

No – see page 87 (top or bottom)

Clinical evidence of vitamin deficiency indicates that the deficiency is of such an order and for such a period that not only are any reserves lost, but the metabolic abnormality has led to cellular changes. Hence clinical evidence represents a relatively late stage in the development of a deficiency disorder.

Action

It is imperative for the appropriate vitamin therapy to be administered as soon as possible, although for some manifestations it is desirable to undertake the necessary confirmatory tests (page 89) and for all it is desirable to establish why the deficiency has occurred.

The main issue normally becomes the question of the administration of a single vitamin in adequate dosage or the use of a multivitamin preparation. Initially there is no doubt that the specific vitamin should be given, for by this means the diagnosis will be confirmed with a therapeutic test. However, once improvement is beginning to occur then the question of additional multivitamin administration requires consideration. Each case should be considered on its merits, taking into account the nature of the disorder and its aetiology. If, for example, there is any suggestion that there is some dietary aspect to the disorder (as opposed, for example, to a specific disorder like pernicious anaemia or a hereditary enzyme state) then it should be assumed that there will be some reduction in the intake of other vitamins and a multivitamin preparation advised in addition to the specific substance.

Is there evidence of low vitamin reserves?

Yes – see below

No – see page 87

The evidence of a low vitamin reserve is likely to follow an epidemiological survey rather than a specific investigation. Under these circumstances the action will depend in large measure on the actual findings. Ideally it would be important to establish whether the reduction is sufficient to interfere with the metabolic activity of the tissues (see page 29 and page 77 *et seq*).

Alternatively, but less commonly, the finding of low vitamin reserves may follow investigation when there is *a priori* evidence suggesting that a vitamin deficiency might be present but clinical evidence is missing. Logically in such circumstances an assessment *should* be made, but in practice this will depend on the facilities that are available and the strength of the *a priori* evidence.

Action

The presence of low tissue reserves implies a negative balance due to low intake, a higher than normal need, or a combination of both.

If the intake is too low advice about a change in the dietary habits should be tried. Whether there should be temporary supplementation to bring the reserves to an appropriate level to enable the new dietary approach to have a chance to sustain them will depend upon a consideration of the particular patient. On balance there are probably merits in so doing if there is any doubt.

If it is impossible to correct the balance by dietary advice, supplementation should be undertaken. While low reserves do not imply a disease at that stage, they require attention, for there are many situations which can lead relatively rapidly from a reduced reserve to an active deficiency with clinical manifestations.

Is there evidence of tissue metabolic abnormality?

Yes – see below

No – see page 88

The finding of tissue metabolic evidence of vitamin inadequacy in the absence of clinical evidence is only likely to arise in the unusual circumstance of a careful research survey, or if a vitamin deficiency is strongly suspected and adequate facilities exist for the appropriate biochemical studies. This is likely to arise in a specialist nutritional situation and this brings it outside the remit of this book. However, it is worth considering the necessary action.

Action

In the circumstances outlined above it is probable that there will be adequate facilities for a clear definition of the reserves and metabolic integrity relating to various vitamins. Hence a clear diagnosis may be made, usually coupled with a clear understanding of the aetiology.

Treatment will, therefore, depend upon a consideration of the aetiology and the circumstances in the individual patient. In general it is likely that specific therapy will be used unless there is clear evidence that the cause implies an imbalance for several of the vitamins and that dietary advice cannot correct the situation (e.g. malabsorption syndromes). Under these latter circumstances appropriate doses of a multivitamin preparation are required.

All tests are negative for vitamin deficiency

The question arises whether under these circumstances there is *any* justification for the administration of one or more of the vitamins. The answer of the purist is quite clear – there is no justification for the administration of the vitamins except for their pharmacological rather than vitamin activity.

Equally it is clear that there will be circumstances in which, despite the clearly negative diagnosis by the practitioner, a confirmatory negative therapeutic response is desirable to satisfy the patient. In these circumstances the practitioner's attention is drawn to the fact that the vitamins have a *very* good safety margin (page 55, with the possible exception of vitamins A and D). Hence a therapeutic trial, while it may produce no benefit is unlikely to do any harm. The author, therefore, finds himself torn between the demands of good scientific medicine and the needs to treat the individual patient, and accepts that there are circumstances in which there are clear merits in showing the patient by a practical therapeutic trial that vitamins are not wonder drugs for all disorders. The author, on occasions has practised what he is here preaching!

14

Laboratory confirmation of vitamin deficiencies

The confirmation of the vitamin nutritional status of an individual or of a group is based upon the use of one or more of the following techniques:

(1) Determination of dietary levels.
(2) Laboratory estimation of the tissue vitamin content.
(3) Measurement of the tissue metabolic status.

Determination of dietary levels

Though the estimation of the dietary intake by indirect or direct means is still a common method for trying to assess the vitamin status of the individual, the results are not very accurate and require very careful interpretation. They take account only of the intake and not the needs. They are more useful for epidemiological purposes than for consideration of the individual.

An account of these methods and an appraisal of their problems and errors is given in Part 1 (page 37) and will not be further considered here.

Laboratory estimation of the tissue vitamin state

The accuracy of a clinical assessment can often be confirmed by laboratory determination of the tissue state. Alternatively a diagnosis may be made or confirmed at a less florid clinical (or even preclinical) stage. This is true for many vitamin deficiency states when access is available to a good laboratory service. In nutritional surveys in the field, however, there are often great practical difficulties to be overcome, both in the collection and the processing of the material.

There are two distinct methods for determining the tissue vitamin status. The first of these is to estimate the degree of body saturation on the basis of circulating blood or plasma levels or by determining urinary excretion. Sensitive and reasonably simple analytical techniques now exist for the determination of the level of many vitamins in the *plasma*. However, the vitamin in the plasma is not filling a metabolic function but is merely in transit from one tissue to another. Thus the plasma status is not necessarily a reliable index of the tissue status. The determination may also be made after a test or loading dose to determine the quantity required to saturate the tissues. However, even this is open to errors as a result of metabolism or urinary loss, for example.

While the daily urinary excretion of the vitamin or its metabolites usually correlates reasonably well with the current dietary intake, it is not always a true reflection of the tissue state, particularly for those vitamins for which there are large tissue reserves. Moreover, the collection of 24 hour samples of urine is tedious. This problem of 24 hour excretion is often overcome by the estimation on a small sample and then correlating the result with creatinine clearance. Although this is easy to perform and widely used for large scale surveys it is not very accurate.

The second method, which involves the estimation of tissue desaturation by direct analysis, is the only method which gives a true indication of the tissue vitamin status. However, measurement of tissue levels is often difficult to undertake due to the problems both of tissue sampling and analytical technique.

Measurement of the tissue metabolic status

All the techniques that have been outlined above seek merely to determine the vitamin level in the body and take no account of the metabolic needs. Tissue vitamin levels only have relevance when equated with the metabolic requirements. To determine the biochemical metabolic efficiency a different group of methods is required.

The tests for biochemical metabolic efficiency attempt to determine the vitamin level of the tissues not in isolation but relative to their metabolic needs. They are based normally on a determination of a metabolic process which utilizes the vitamin as coenzyme. Unfortunately, reliable tests are available for only relatively few vitamins. The technique used should be specific to the vitamin; there should be no alternative paths or coenzymes; ideally the metabolic stage should be rate-limiting if the result is to have relevance to the integrity of the metabolic process as a whole.

No single method (dietary determination; plasma or tissue concentration; tissue metabolic status) gives reliable results under all circumstances. Any result of the determination of the vitamin status by any of these techniques requires careful interpretation. Nevertheless, used with care they can

Table 37 Laboratory methods for determining tissue saturation and tissue metabolic integrity

Vitamin	Serum or blood level	Urinary exretion	Tissue determination	Metabolic test
Retinol	serum retinol	—	—	dark adaptation
Calciferol	serum 25-hydroxycholecalciferol or 1,25-dihydroxycholecalciferol	—	—	serum alkaline phosphatase bone radiography
Tocopherol	serum tocopherol	—	—	urinary creatinine excretion red cell peroxide hydrolysis
Phylloquinone	serum phylloquinone	—	—	specific clotting factor determination
Thiamine	plasma thiamine	thiamine	—	raised pyruvic acid and lactate/pyruvate ratio and 2-oxoglutarate in blood, red cell transketolase, TTP effect
Riboflavine	erythrocyte riboflavine	riboflavine	—	erythrocyte glutathione reductase, FAD effect
Pyridoxine	blood pyridoxal phosphate	4-pyridoxic acid	—	tryptophan load test, erythrocyte aspartate aminotransferase and glutamate oxaloacetate transaminase
Niacin	—	methyl nicotinamide	—	—

91

Table 37 Laboratory methods for determining tissue saturation and tissue metabolic integrity

Folic acid	serum or red cell folic acid	—	bone marrow and blood picture, histidine load test and deoxyuridine test
Cobalamin	plasma cobalamin	cobalamin absorption test	bone marrow and blood picture, deoxyuridine test, urinary methylmalonic acid, achlorhydria
Pantothenic acid	blood pantothenic acid	—	acetylation activity
Biotin	plasma biotin	biotin absorption test	—
Ascorbic acid	plasma ascorbic acid	ascorbic acid	buffy coat or leukocyte ascorbic acid intradermal dye test	—

provide very valuable additional evidence beyond that which is available from clinical examination. Specifically they can indicate the existence of possible problems before they produce clinical disorders.

A summary of all these laboratory methods that can be applied to the individual vitamins is given in Table 37. A detailed description of their performance is outside the re,nit of this book, but will be found in the large textbooks on nutrition.

Guidelines for acceptable and definitely deficient levels in common laboratory determinations are given in Table 38. Caution should be exercised over marginal figures for different methods of determination may give somewhat different levels.

Table 38 Guidelines for consideration of inadequate vitamin status by laboratory studies. The values quoted are those for adults

Nutrient and units		Definitely deficient	Acceptable	Possible vitamin disorder
Haemoglobin (gm/dl)	male	up to 12.0	14.0 +	vitamin B$_{12}$
	female	up to 10.0	12.0 +	folic acid
				vitamin C
Haematocrit (%)	male	up to 37	44 +	vitamin B$_{12}$
	female	up to 31	38 +	folic acid
				vitamin C
Serum ascorbic acid (mg/dl)		up to 0.1	0.2 +	vitamin C
Plasma vitamin A (µg/dl)		up to 10	20 +	vitamin A
Plasma carotene (µg/dl)		up to 20	40 +	vitamin A
Serum folacin (ng/ml)		up to 2.0	6.0 +	folic acid
Serum vitamin B$_{12}$ (pg/ml)		up to 100	100 +	vitamin B$_{12}$
Thiamine in urine (µg/g creatinine)		up to 27	65 +	vitamin B$_1$
RBC transketolase-TPP-effect (ratio)		25 +	up to 15	vitamin B$_1$
Riboflavine in urine (µg/g creatinine)		up to 27	80 +	vitamin B$_2$
RBC glutathione reductase-FAD-effect (ratio)		1.2 +	up to 1.2	vitamin B$_2$
Tryptophan load (mg zanthurenic acid excreted)		25 + (6 h)	up to 25	niacin
Urinary N'-methyl nicotinamide (mg/g creatinine)		up to 0.2	0.6 +	niacin
Urinary pyridoxine (µg/g creatinine)		up to 20	20 +	vitamin B$_6$
Urinary pantothenic acid (µg)		up to 200	200 +	pantothenic acid
Plasma vitamin E (mg/dl)		up to 0.2	0.6 +	vitamin E

15

Therapeutic index

Therapeutic indications can be divided into broad classes.

First, the vitamins can be used as replacements for established primary or secondary vitamin deficiency states. There is no doubt about their value under these circumstances. Second, there are several indications where there is no vitamin deficiency but the therapeutic value of the vitamin has been confirmed by adequately controlled studies. Third, there are disorders for which there is reasonable anecdotal evidence of efficacy. In these disorders use is, in the opinion of the author, fully justified when other treatment is inappropriate or ineffective.

A bibliography of the key papers is given for the therapeutic index (page 199).

Acne. High oral doses of vitamin A (100 000 – 200 000 i.u.) and topical retinoic acid have been used successfully. Retinoids are used in *specialist centres only for severe cases.*

Acute schizophrenia. See schizophrenia.

Alcoholism. The main vitamin deficiency encountered in alcoholism is that of thiamine which should be given at a dose of 100 – 300 mg daily. However, there is frequently a general nutritional deficiency in chronic alcoholism including folic acid, vitamin B_6, niacin, riboflavine and pantothenic acid and a multivitamin preparation should also be given at high dosage.

Angular stomatitis. This is usually due to a riboflavine deficiency, but since there is frequently a multiple vitamin deficiency it is convenient to administer a multivitamin preparation.

Anosmia. In some cases where atrophic rhinitis appears to be the cause, high doses of vitamin A (100 000 – 200 000 i.u.) daily have been used with success.

Antibiotics. Several antibiotics, particularly those of broad-spectrum type, can lead to a deficiency of various vitamins particularly those of the B complex. A multivitamin preparation should be given prophylactically with broad-spectrum antibiotics.

Anticoagulant bleeding. This can be reversed with vitamin K_1 (5−10 mg parenterally).

Atrophic rhinitis. High doses of vitamin A (100 000−200 000 i.u.) daily have been used with success in some cases.

Beefy-red tongue. Responds rapidly to niacin (50−100 mg daily).

Beri-beri. This is a manifestation of thiamine deficiency which should be given at a daily dose of 50−200 mg. Since there is usually a more generalized vitamin deficiency a multivitamin preparation should also be given.

Biliary surgery. Administer vitamin K_1 (5−10 mg i.m.) prophylatically.

Bitot's spots. This may be an early sign of hypovitaminosis A and should be treated with this vitamin (10 000−50 000 i.u. daily).

Bow legs. This may be a sign of rickets and should be investigated and treated accordingly.

Broad-spectrum antibiotics. See antibiotics.

Burning feet syndrome. A manifestation of pantothenic acid deficiency which responds to a dose of 100−200 mg daily.

Carpal tunnel syndrome. In some patients it may be associated with pyridoxine deficiency and a dose of 20−100 mg daily may be tried.

Cerebral beri-beri. See Wernicke's encephalopathy.

Cerebral haemorrhage. There is a suggestion that cerebral haemorrhage in young subjects correlates with low vitamin C levels, and vitamin C levels should be maintained to reduce the incidence. Intracerebral haemorrhage in premature babies is reduced by vitamin E administration.

Cheilosis. This is usually due to a deficiency of riboflavine which should be treated preferably with a multivitamin preparation since deficiencies of other vitamins may also be present.

Chinese restaurant syndrome. This reaction to high levels of monosodium glutamate is reputed to be eased by pyridoxine (50 mg).

Coeliac disease. See malabsorption syndrome.

Cirrhosis. This may lead to low vitamin K levels and a bleeding diathesis. Vitamin K_1 should be administered, particularly if surgery is contemplated.

Common cold. Although there is no conclusive evidence of the value of vitamin C in the common cold, the author is one of the many doctors who advise its use in high doses (1–2 g daily) in the early stages of the infection.

Convulsions in infancy. See infancy convulsions.

Coryza. See common cold.

Cramps. These may respond to riboflavine (20 mg daily).

Craniotables. Vitamin D (5000 i.u. daily) should be administered.

Cystine stones. This is a rare disorder for which vitamin C can be used prophylactically.

Dark adaptation reduction. See night blindness.

Dementia. This is a classical manifestation of pellagra and in such patients it responds to niacin (50–100 mg daily). It may also occur in folic acid and vitamin B_{12} deficiency.

Depression. When this is part of premenstrual tension or occurs in coeliac disease, pyridoxine (50–200 mg daily) may be effective.

Dermatitis. If this is due to pellagra it responds rapidly to niacin (50–100 mg daily).

Desquamative erythroderma. See Leiner's disease

Diarrhoea. The presence of diarrhoea can lead to various vitamin deficiencies depending upon the cause. If there is steatorrhoea the fat soluble vitamins and folic acid are very likely to be deficient and should be given. Diarrhoea is a fairly common effect in the early stages of high vitamin C dosage. It may also be due to pellagra and responds to niacin (50–100 mg daily).

Dry beri-beri. See beri-beri.

Dupuytren's contracture. This may respond to high doses of vitamin E (200–400 mg daily).

Encephalomalacia. This may be a rare manifestation of vitamin E deficiency in infants (page 134). If the diagnosis is confirmed high vitamin E dosage is required.

Familial cystathioninuria. This responds to pyridoxine (50–100 mg daily).

Familial xanthurenic acid disease. This responds to pyridoxine (50–100 mg daily).

Fat transport diseases. The fat soluble vitamins are usually depleted in these disorders and should be supplemented. This includes particularly vitamin E.

Geographical tongue. This may be associated with ariboflavinosis and the possibility should be studied.

Glossitis. If this is due to pellagra it responds rapidly to niacin (50–100 mg daily).

Haemodialysis. This leads to low levels of the water soluble vitamins and raised levels of those that are fat soluble. Supplementation with B group and vitamin C is desirable.

Haemolytic anaemia. Rarely, in prematures a haemolytic anaemia may occur in vitamin E deficiency and should be treated (5–25 mg daily). This is also a rare adverse effect to menadione in infancy and an exchange transfusion may be necessary.

Haemorrhagic disease. See neonatal haemorrhagic disease.

Haemosiderosis. In this disorder there is an excess gastrointestinal absorption of iron, and the administration of ascorbic acid which facilitates absorption should be avoided.

Hartnup disease. This familial pellagra-like disorder may be treated with niacin (50–100 mg daily) with benefit.

Hyperbilirubinaemia. In infancy this may be a rare complication of menadione administration and an exchange transfusion may be necessary. In adults the presence of high bilirubin levels is an indication for the administration of vitamin K_1, particularly before surgery.

Hypercarotinaemia. This is a rare disorder due to too high an intake of carotene. There are no adverse effects but the intake of carotene should be discontinued.

Hypercholesterolaemia. One of the methods that has been used for the treatment of this disorder is nicotinic acid (1000–3000 mg daily).

Hyperemesis gravidarum. This may be treated with pyridoxine (100–200 mg daily).

Hyperkeratotic hair follicles. Should be treated with vitamin C (500–1000 mg daily).

Hypervitaminosis A. This is considered in detail on page 119. The administration of vitamin A should be discontinued immediately.

Hypervitaminosis D. This is one of the few worrying problems of high vitamin dosage (page 131), particularly in infants. Further vitamin D should be withheld immediately.

Hypoparathyroidism. This may be treated with high doses of vitamin D judging the correct dose by the plasma calcium levels. The normal maintenance dose is in the range 50 000–200 000 i.u. daily.

Idiopathic steatorrhoea. Among the vitamin deficiencies seen in this disorder are the fat soluble members and folic acid.

Infancy convulsions. These may be due to pyridoxine deficiency and if so can be relieved by pyridoxine (up to 100 mg per day).

Infantile seborrhoeic dermatitis. A sign of biotin deficiency. A daily dose of 2−5 mg should be given.

Infections. There is clear evidence of vitamin C depletion during the course of an infection. Vitamin C should be administered in quantities sufficient to restore and maintain adequate vitamin C reserves (500 mg daily).

Intermittent claudication. This may respond to high doses of vitamin E (400−600 mg daily).

Irradiation sickness. Pyridoxine may help to relieve this (100−200 mg daily).

Isoniazid neuritis. This should be treated with pyridoxine (100−200 mg daily) and isoniazid discontinued.

Keratomalacia. This responds to vitamin A (daily dose 10 000−50 000 i.u.).

Kernicterus. This is a rare but serious adverse effect of menadione administration in infants. Exchange transfusion may be necessary.

Korsakoff's psychosis. A manifestation of alcoholism in which hypovitaminosis B_1 may be one factor. High doses (100−300 mg daily) should be given, and since there is frequently a multiple vitamin deficiency a multivitamin preparation may be given in addition with benefit.

Leiner's disease. Should be treated with biotin (2−5 mg daily).

Leukopaenia. Several vitamin deficiencies can lead to leukopaenia. The most common are folic acid and vitamin B_{12}.

Liver damage. This is a rare manifestation of overdose of nicotinamide or vitamin A and requires withdrawal of the vitamin. The presence of liver damage may lead to vitamin K deficiency, and vitamin K_1 should be administered particularly if any surgery is contemplated.

Macrocytic anaemia. This may be due to either folic acid or vitamin B_{12} deficiency (see pages 167 and 173). If there is any doubt about the cause use vitamin B_{12} rather than folic acid, to avoid precipitating subacute combined degeneration of the cord.

Malabsorption syndrome. Intestinal malabsorption leads to the deficiency of many vitamins and should be treated with high doses of a multivitamin preparation including fat soluble members.

Maple syrup disease. This may be improved with the administration of vitamin B_1 (daily doses of 50−200 mg).

Megaloblastic bone marrow. This may be due to either folic acid or vitamin B_{12} deficiency. Since the administration of folic acid to a pernicious anaemia (vitamin B_{12} deficiency) patient may precipitate subacute combined degeneration of the cord, if there is any doubt about the cause, vitamin B_{12} should be given.

Mental retardation. There are many vitamin deficiency causes for this. These include Wernicke's encephalopathy *q.v.*, pellagra *q.v.* and familial cystath-ioninuria *q.v.*

Migraine. It is reputed that some cases respond to the vasodilator effect of nicotinic acid (100 mg).

Nasolabial dermatitis. Frequently due to riboflavine deficiency. Since there is often a multivitamin deficiency a multivitamin preparation may be given with benefit.

Neonatal haemorrhagic disease. Administer 0.5–1.0 mg vitamin K_1 intramus-cularly. This dose should be given prophylactically to the baby or a higher dose (5 mg i.m.) administered to the mother just before the delivery.

Neonatal retinitis. This may occur with high oxygen levels and hypovitamin-osis E. High levels of oxygen should be avoided and vitamin E (5–25 mg daily) should be given.

Neural tube defects. There is evidence that these may be the result of vitamin deficiency (perhaps folic acid). Since the exact cause is unknown there are merits in giving those who are at risk a multivitamin supplement during pregnancy with a folic acid intake of at least 10 mg per day.

Neuritis on isoniazid. See isoniazid neuritis.

Neurolathyrism. This is a rare disorder which occurs in India and Bangladesh due to the consumption of *L.sativa* (khesari) in times of drought. Adequate vitamin C status prevents the disorder and vitamin C (500–1000 mg daily) can cure the developed paralysis.

Neuropathy. A common cause of a neuropathy, particularly in alcoholics or the elderly is a thiamine deficiency, and this should be treated with thiamine at a dose of 50–200 mg daily. Neuropathy is a rare complication of dramatically high pyridoxine dosage (2–6 g per day for years) and requires discontinuation of the vitamin.

Night blindness. This is a manifestation of vitamin A deficiency and should be treated with 10 000–50 000 i.u. daily.

Normocytic anaemia. This may rarely be due to either a riboflavine or pyridoxine deficiency and high doses of these vitamins may be tried if other causes are excluded.

Osteomalacia. This should be treated with vitamin D (5000–10 000 i.u. daily) together with calcium.

Otosclerosis. This may be due to vitamin A deficiency and should be treated with doses of 10 000–50 000 i.u. daily.

Paralytic ileus. Parenteral pantothenic acid (200–600 mg) may be used as part of the therapeutic regimen.

Pellagra. Should be treated with niacin (50–100 mg daily).

Pellagra sine pellagra. A mental disorder without the other signs of pellagra which should be treated with niacin (50–100 mg daily).

Peyronie's disease. This may respond sometimes to high doses of vitamin E (200–600 mg daily).

Photodermatitis. This may be a manifestation of pellagra and responds to niacin (50–100 mg daily).

Photophobia. This may be due to ariboflavinosis and this possibility should be considered.

Polyneuritis. This is often due to a thiamine deficiency and vitamin B_1 should be given (50–200 mg daily).

Polyunsaturated fatty acids. There is evidence that high levels of intake of these can lead to a vitamin E deficiency and a supplement of 50–100 mg daily should be administered.

Post gastrectomy. The post gastrectomy patient may suffer from various vitamin deficiencies including riboflavine. A multivitamin preparation may be administered with benefit.

Pregnancy cramps. See cramps in pregnancy.

Premenstrual tension. Pyridoxine (50–75 mg twice a day) started 3 days before the expected start of the period may help to relieve this.

Psychosis. This may rarely be the result of folic acid or vitamin B_{12} deficiency.

PUFA. See polyunsaturated fatty acids.

Rachitic rosary. This is due to vitamin D deficiency in infancy and should be treated with doses of 5000–10 000 i.u. daily depending on the weight and severity.

Raw fish neuropathy. A disorder of the Far East (particularly Japan). Thiamine (50–200 mg daily) should be given.

Renal rickets. This should be treated with calcitriol.

Resistent rickets. This is a general term covering a number of disorders (*see*

Table 45, page 129). In general calcitriol is the substance of choice but such cases require specialist attention.

Retrolental fibroplasia. This reaction, which is due to too high a level of oxygen administration in babies, may be reduced by the administration of vitamin E (5–25 mg daily) to the neonate.

Rickets. Administer 5000 i.u. vitamin D daily in infants, rising to 10 000 i.u. daily in severe cases and 15 000 i.u. in late onset cases. Alternatively, 200 000–400 000 i.u. can be given as a single oral or intramuscular dose.

Schizophrenia. There was a period during which acute schizophrenia was treated with high doses of niacin. However, with the use of modern neuroleptic drugs the use of niacin has declined.

Scleral vascularization. This may be due to riboflavine deficiency and the possibility should be investigated.

Scrotal dermatitis. This may be due to riboflavine deficiency and this should be investigated.

Scurvy. This is due to ascorbic acid deficiency and should be treated with 500–1000 mg per day until the tissue reserves are saturated.

Senile osteoporosis. Vitamin D deficiency is one factor, and adequate vitamin D and calcium should be given prophylactically in the elderly.

Skin flushing. This is a common manifestation of nicotinic acid administration and may be avoided by the use of nicotinamide.

Skin keratinization. See toad skin.

Skull bossing. This is a sign of vitamin D deficiency in infancy and vitamin D (5000–10 000 i.u. daily) should be given.

Smoking. There is clear evidence that smoking increases vitamin C requirements by about 40%, and in smokers a higher vitamin C intake is desirable.

Spasmophylia. This may be precipitated by vitamin D in infants and younger children. To avoid it, vitamin D should be supplemented by the simultaneous administration of oral calcium.

Stomatitis. This may be due to niacin deficiency and should be treated with niacin at a dose of 50–100 mg daily.

Subacute combined degeneration. This is a manifestation of vitamin B_{12} deficiency. Doses of 10–20 μg per day should be given. Folic acid is contraindicated since it makes the nerve degeneration worse.

Surgery. Ascorbic acid can be given with benefit to wound healing if a deficiency is suspected. In surgery of disorders of the biliary tract or liver there are merits in prophylactic vitamin K_1 (10 mg).

Thiamine hypersensitivity. A hypersensitivity reaction is an uncommon feature of parenteral thiamine administration. A skin test dose should be given before the use of parenteral thiamine.

Toad skin. This was previously thought to be due to hypovitaminosis A but is now thought to be due to essential fatty acids deficiency. However, vitamin A (10 000 – 50 000 i.u. daily) may be worth a trial in resistant cases.

Trigeminal neuralgia. On occasions this may respond dramatically to high doses of vitamin B_1 (200 – 600 mg daily).

Urticaria. This may occur in familial xanthurenic acid disease and responds to pyridoxine (50 – 100 mg daily).

Vulval dermatitis. This may be a manifestation of riboflavine deficiency and should be investigated appropriately.

Wernicke's encephalopathy. This is the cerebral form of beri-beri and should be treated with high doses of vitamin B_1 (100 – 400 mg daily). A condition similar to this may also be seen in pellagra, which responds to niacin (50 – 100 mg daily).

Wet beri-beri. See beri-beri.

Xanthurenic acid disease. See familial xanthurenic acid disease.

Xerophthalmia. Daily doses of vitamin A are required (10 000 – 50 000 i.u.).

Xerosis. See xerophthalmia.

16

Interactions between drugs and vitamins

Many of the drugs in current use depend for their effects on the selective inhibition of various enzyme systems. Since the vitamins are themselves often involved as cofactors in enzyme action, it follows that interaction is likely. This involves not only drugs altering the vitamin status but the vitamin economy altering the effects of drugs.

As shown in Table 39, drugs may affect vitamin activity in many ways, and this table also gives examples of some of the interactions that occur at the various sites.

Table 39 Examples of effects caused by drug vitamin interactions

Effect	Drug(s)	Vitamins involved
Decreased appetite	amphetamines	all
Impaired absorption	neomycin	vitamins A, D, K, B_{12}
	cytotoxic drugs	all
Changed distribution	salicylates	folic acid, vitamin C
Interference at enzyme or receptor	coumarins	vitamin K_1
	methotrexate	folic acid
Changed metabolism or excretion	oral contraceptives	folic acid, vitamins B_6, B_{12}
	barbiturates	vitamins D, B_{12}, folic acid

A summary of the main drug–vitamin interactions, subdivided according to the vitamins, is given in Table 40.

Table 40 Some of the drug–vitamin interactions reported in the literature

Vitamin	Drugs with which vitamin interacts
A	cholestyramine, corticosteroids, mineral oil, neomycin, bleomycin, vinblastine, benzpyrene, spironolactone, DDT
D	cholestyramine, anticonvulsants, barbiturates, glutethemide, mineral oil, neomycin, bleomycin
E	cholestyramine, mineral oil, oxygen
K	cholestyramine, oral anticoagulants, mineral oil, salicylates, antibiotics including neomycin, cephalosporins
B_1	oral contraceptives, thio semicarbazone, 5-fluouracil
B_2	oral contraceptives, galactoflavine, boric acid
B_6	levodopa, oral contraceptives, hydralazine, isoniazid, penicillamine, 4-deoxypyridoxine, thio semicarbazone, cycloserine
Niacin	isoniazid, 6-amino nicotinamide, benzerazide, carbidopa
B_{12}	potassium chloride, cholestyramine, colchicine, oral contraceptives, neomycin, bleomycin, P-amino salicylic acid, biguanides, anticonvulsants, metformin, phenformin
Folic acid	anticonvulsants, barbiturates, cholestyramine, oral contraceptives, methotrexate, pyrimethamine, salicylates, triamterine, trimethoprim, antacids
Pantothenate	thio semicarbazone
Vitamin C	smoking, oral contraceptives, salicylates, tetracycline, corticosteroids, calcitonin, alcohol

While it is usual to be mainly concerned with the undesirable effects that drugs may exert on the vitamin status, it is known that the vitamin level can also on occasion affect drug activity. One specific effect is exerted on the hepatic microsomal mixed function oxidases which are responsible for one of the two main pathways for drug metabolism. Various vitamin deficiencies, e.g. vitamins A, E, C and possibly B_2 can reduce the activity in these enzymes. This can be demonstrated experimentally in humans by measuring the half-life of antipyrine, 95% of which is metabolized by these oxidases.

While this activity means that the presence of adequate vitamin levels reduce the half-life there are a few circumstances where increases in vitamins increase drug actuvity. These are summarized in Table 41.

Table 41   Some of the interactions in which the vitamin level can alter the drug activity

Vitamin	Change	Drug	Effect
Vitamin A	increase increase	cyclophosphamide chlorambucil	increases effect increases effect
Vitamin K$_1$	increase	oral anticoagulants	decreases effect
Thiamine	increase increase decrease	thiosemicarbazone zoxozolanine heptachlor aniline	increases effect decreased metabolism increased metabolism
Vitamin B$_2$	decrease decrease	benzpyrene aminopyrine hexobar- bital aniline	decreased metabolism increased metabolism
Vitamin B$_6$	increase	levodopa	decreases effect
Vitamin C	increase increase	ethyl oestradiol oral anticoagulants	increases effect decreases effect

PART 3

THE INDIVIDUAL VITAMINS

17

Vitamin A

Alternative names: retinol; axerophthol

The term vitamin A covers a group of fat soluble compounds which occur in several isomeric forms and have the qualitative activity of retinol. Of these the physiologically most active forms, and those that are found in mammalian tissues, are *all trans* vitamin A and the related 11-*cis* vitamin A (Figure 9). Naturally occurring vitamin A is only present in the animal kingdom, although vegetables contain carotenoids which are precursors of vitamin A. Vitamin A is often found in nature in an esterified form as the acetate or the palmitate.

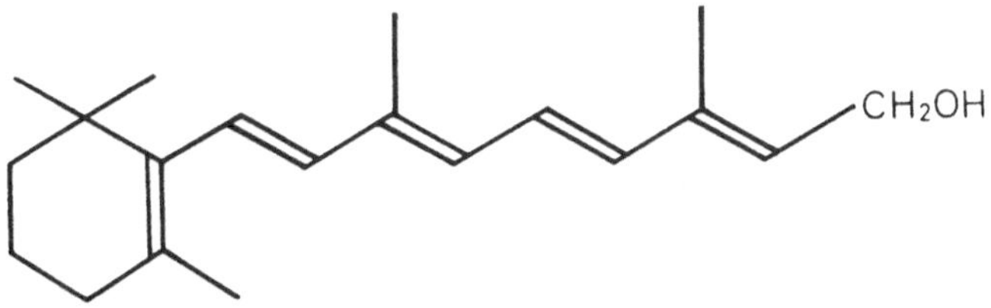

Figure 9   Structural formula of retinol

Sources

A proportion of the daily intake comes as preformed vitamin A from animal sources, but the remainder is taken in as one of the carotenoid provitamins from the vegetable kingdom. The various carotenoids form part of the yellow and orange pigments of most fruits and vegetables. The β-carotene is the form which is most readily converted to vitamin A. The conversion is mainly achieved by terminal oxidation.

The main dietary intake in most industrialized countries is currently

111

derived from dairy products and margarine, the latter being supplemented with vitamin A in many countries.

Requirements

The daily requirements for humans are summarized in Table 7. The daily intake is now expressed as 'μg retinol equivalents' (RE), i.e. vitamin A is expressed by weight, and the carotenoids the weight of vitamin A to which they would be converted in the body, 1 μg of retinol is equivalent to 3.33 international units of vitamin A. There is a fairly narrow range of the requirement levels defined by the health authorities of different countries with a mean for adult males for 1.0 mg RE. A supplement of 0.2 mg is advised during pregnancy and 0.4 mg during lactation.

About 75% of the requirement is covered by preformed vitamin A in the diet, the remaining 25% being met from the conversion of carotenoids. At levels corresponding to the normal intake of carotenoids, the conversion factor is 6 μg β-carotene or 12 μg of other carotenoids with a provitamin A action, corresponding to 1 μg of retinol.

Causes of deficiency

The main mechanisms that lead to the production of a vitamin A deficiency are shown in Figure 10. In practice it is common for vitamin A deficiency to occur as part of an overall state of general malnutrition. In these circumstances the vitamin A deficiency may be partly due to the inadequate level in the diet and in part to low levels of transport proteins.

Recent surveys have indicated that vitamin A deficiency is widespread, particularly in children (Table 42).

Assessment of vitamin status

Vitamin A status may be determined by either the serum retinol level or by measuring dark adaptation.

Activity

The carotenoids present in the diet – particularly β-carotene – can be transformed within the intestinal tract into vitamin A. There is active absorption of vitamin A present in the diet in the form of retinyl esters or vitamin A derived from carotenoids, probably with a low-density lipoprotein carrier. Absorption is helped by the presence of bile salts as emulsifying agents. The vitamin is stored in the Kupffer cells of the liver, normally in the form of fatty acid esters, and is subsequently carried in the blood bound to a prealbumin – vitamin A binding protein. Vitamin A is used

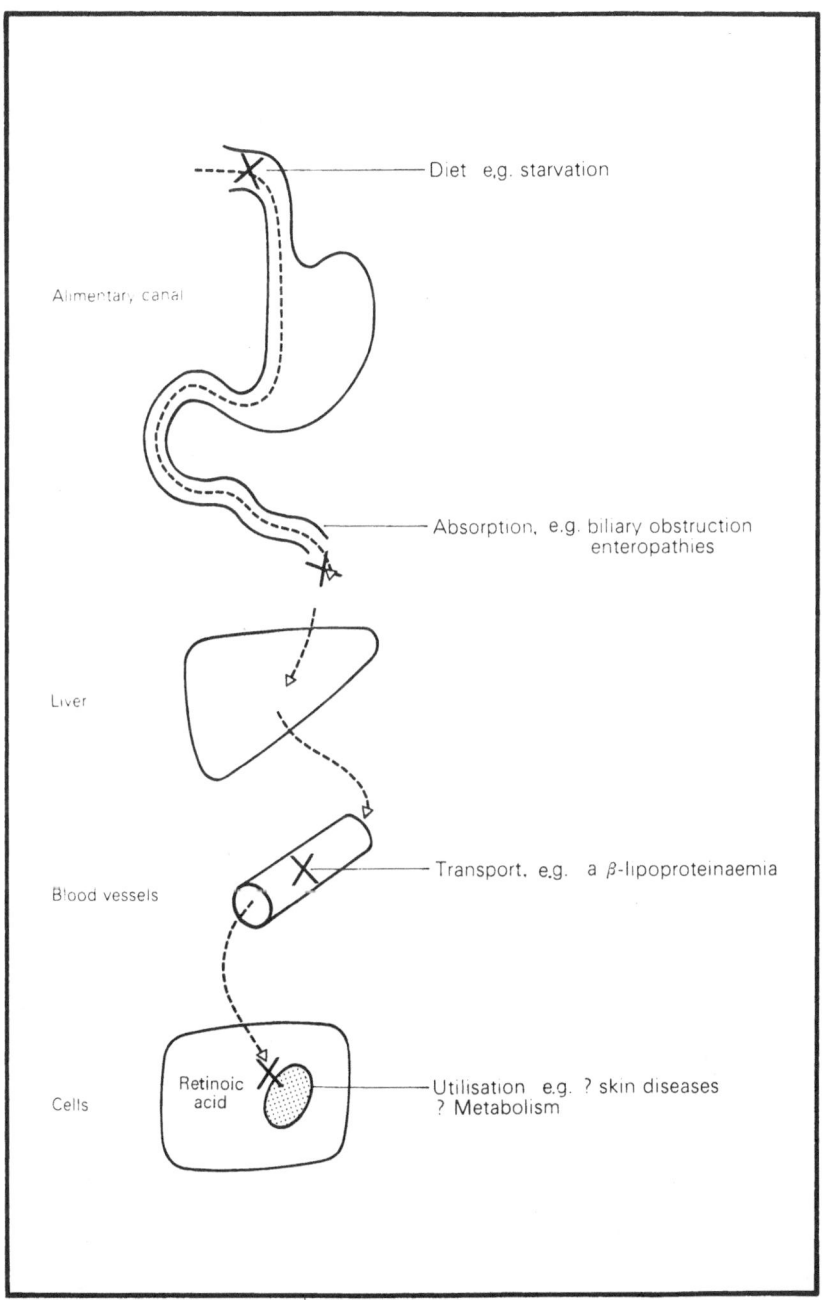

Diet e.g. starvation

Alimentary canal

Absorption, e.g. biliary obstruction
enteropathies

Liver

Transport. e.g. a β-lipoproteinaemia

Blood vessels

Retinoic
acid

Utilisation e.g. ? skin diseases
? Metabolism

Cells

Figure 10 Main mechanisms for the production of hypovitaminosis A

113

Table 42 Recent surveys demonstrating vitamin A malnutrition in several countries

Country	Population group (number)	Comment	Ref.
India	Preschool children	Ocular signs of vitamin A deficiency in 5–10% of children	Reddy, 1980
Philippines	Children aged 1–16 (1715)	4.5% with clinical and 57% with bio-chemical signs	Solon et al., 1978
Central America	Children aged 0–14 (1898 families)	24% children with vitamin A serum levels below 20 µg	Arroyave et al., 1979
Indonesia	Children	More than 60 000 children develop corneal involvement; at least one third are left permanently blind or visually impaired	Tarwotjo, 1980
Columbia	Population survey (52 762)	12.4% showed low levels	Mora (in press)
USA	Black males and females aged 12–34 years	High proportion with follicular hyperkeratosis	Loewenstein, 1981
Brazil	Preschool survey in various parts of Brazil	Xerosis in up to 30% preschool children Between 40 & 70% with low plasma levels	Wilson et al., 1983
Central American	Surveys in different countries Children aged 0–14 years	Intake < 100% RDA Clinical xerosis	INCAP, 1972
Costa Rica (414)		88% 23.3%	
El Salvador (278)		98% 36.5%	
Guatemala (200)		83% 17.8%	
Honduras (323)		94% 30.3%	
Nicaragua (331)		89% 14.9%	
Panama (352)		92% 13.4%	
Brazil	Food & population survey	Consumption varies between 10–62% RDA; 24% showed low plasma level	Roncado, 1980
USA	General survey age 1–74	Abnormal levels in 1–10% of groups	DHEW 1979

in the eye as the related aldehyde (retinal) where it has a specific action in rod and cone vision. In many peripheral tissues vitamin A has a more general metabolic effect, and in certain cells, particularly those of the epithelium, it appears to be irreversibly converted into the active retinoic acid.

The peripheral effect is still poorly understood. The only known metabolic actions are via 'active sulphate' in the synthesis of mucopolysaccharides and in the synthesis of corticosteroids. It alters the stability of the membranes of mitochondria and lysosomes, but the relationship of these effects to the pathological signs of deficiency, mainly manifest in a loss of membrane integrity is far from clear. Vitamin A is also necessary for spermatogenesis, oogenesis, placental development and embryonic growth.

Whether vitamin A or the precursor β-carotene can protect against the development of cancer is currently a matter of dispute, although recent studies suggest that such an activity may be present.

The visual aspect is specific to vitamin A. That which has been most throughly investigated is concerned with scotopic (dim-light) vision by the rods (Figure 11). The rods contain a pigment (rhodopsin) consisting of a specific protein, photopsin, bound to the 11-*cis* isomer of retinal (i.e. the aldehyde form, chemically designated retinene) as a chromophore.

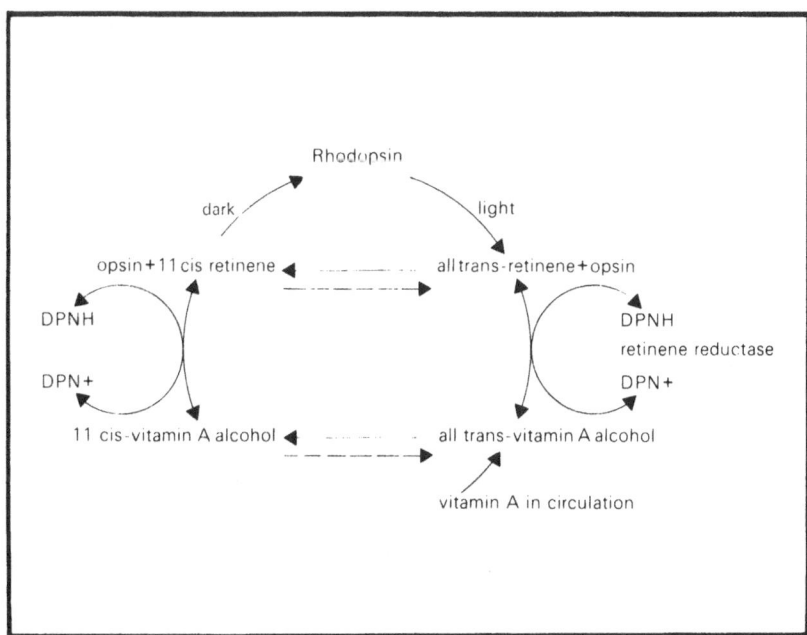

Figure 11 Biochemical mechanism for the role of vitamin A in rod vision

Exposure to light changes the 11-*cis* form to the all-*trans* retinal which no longer forms a stable complex with photopsin. The generation of energy in the optic nerves which is interpreted as scotopic sight is the result of the exposure of active groups on the released opsin. It may depend on the action of a 'second messenger', perhaps calcium ions. In the dark the all *trans* retinal is either isomerized directly to 11-*cis* retinal again or reduced to all *trans* vitamin A, isomerized to 11-*cis* vitamin A and then oxidized to the aldehyde form before recomplexing with opsin to reform the active rhodopsin.

The cones contain three rhodopsins with different opsins, but the same chromophore 11-*cis* retinal. These are sensitive to different wavelength bands covering a broad spectrum of colour. They only respond when the light has a higher intensity than that which stimulates the rods.

Vitamin A is also found in high concentrations in other sensory receptors (Figure 13) and vitamin A deficiency may be associated with reduced sensitivity of these receptors. Hence it has been suggested that this vitamin or a metabolite may be an important constituent of receptors for most of the special senses.

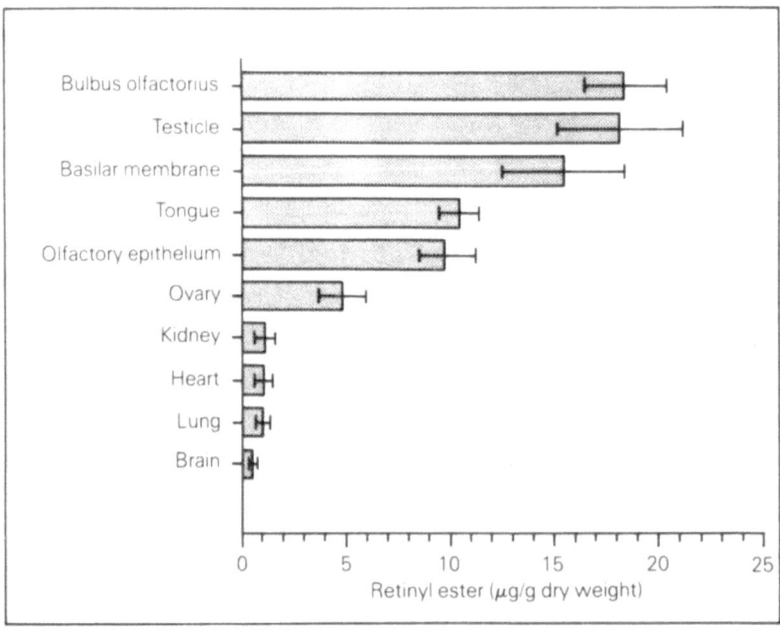

Figure 12 Retinyl ester content of various tissues of the body. (Based on Biesalski, Cartagena symposium, 1984)

Clinical manifestations of deficiency

In clinical practice vitamin A deficiency is commonly associated with general malnutrition and intercurrent infections which complicate the picture.

The vitamin A level may have an influence on the outcome of intercurrent infections in children. Thus in one recent study in Indonesia the mortality rate correlated well with predetermined clinical evidence of retinol deficiency (Figure 13), and this still applied when respiratory infections, diarrhoea and the degree of wasting was taken into account. The exact relationship of vitamin A to the reduction of childhood mortality still needs confirmation.

The only current *unequivocal* signs of vitamin A deficiency occur in the eyes. Depletion of the retinene leads to poor dark adaptation, which shows a close correlation with the blood level of retinol and is readily corrected by therapy.

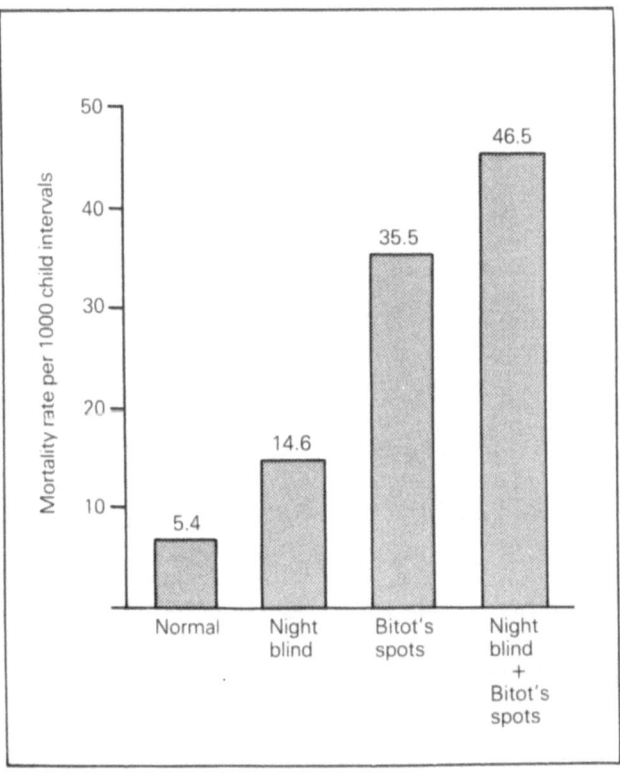

Figure 13   The relationship of mortality to evidence of vitimin A depletion in children. (Based on Sommer et al., Lancet, **2**, 585 (1983))

The eye manifestion of a more serious deficiency is xerophthalmia. This is a drying and degenerative disease of the cornea in which the exact signs depend not only on the severity of the deficiency but also the age of the patient and concurrent local infection. Two main stages are usually described. In xerosis (Plate 5), there is dryness of the conjunctiva and a wrinkled thickened appearance: the relationship of Bitot's spots (white pearly plaques – Plate 6) to xerosis is still far from clear. In more severe cases, and particularly if the eye becomes infected, xerosis can progress to keratomalacia with softening of the cornea, involvement of the iris and lens and ultimately extensive scarring leading to blindness (Plate 7).

Although it is rare to find either xerosis or xerophthalmia in industrially developed countries it is common elsewhere, particularly in the Middle East, and is still among the most common of all causes of blindness worldwide.

Although very pronounced skin keratinization (toad skin) was formerly attributed to vitamin A deficiency it is now usually ascribed to deficiency of the essential fatty acids. Nevertheless, there is presumptive evidence from animal experiments and human clinical observations, that vitamin A deficiency leads to some keratinization of both the skin and squamous mucosal surfaces with loss of surface integrity and increased susceptibility to infections.

Therapy

Deficiency states

Deficiency states should be treated as soon as they are diagnosed with relatively high doses of vitamin A (a daily dose of 20 000 – 50 000 i.u. daily is normally appropriate). When minor grades of vitamin A deficiency are suspected (e.g. in malabsorption syndrome, gastrectomy, etc.) doses from 5000 – 30 000 i.u. daily are normally regarded as adequate.

There has been recent interest in the use of very high doses (of the order of 300 000 i.u. of oil miscible vitamin A at intervals of either 6 months or 1 year for the prophylaxis of deficiency states in areas where hypovitaminosis is endemic.

Note should be taken of the safety margin of vitamin A which is less than that for most other vitamins (see below). Dosage should always take account of the possibility of hypervitaminosis A. Vitamin A should be used very cautiously during pregnancy.

Other disorders

In addition to the use of vitamin A in deficiency states, high doses (100 000 – 200 000 i.u.) of this vitamin have been administered in a wide range of dermatological and epithelial disorders, including acne, atrophic rhinitis, anosmia, tinnitus and otosclerosis, though with rather mixed results. Topical retinoic acid has also been used for certain skin disorders, but for such disorders the retinoids are now preferred (page 120 *et seq.*) though their *toxicity currently precludes their use other than in specialist hospital practice.*

Safety

Adverse health effects have been reported in people who ingest excessive quantities of vitamin A and these are associated with high serum and liver vitamin A levels (some 10 times the normal). The health status is particularly important in the consideration of safety. Thus for example in those with liver disease the level of unbound plasma vitamin A, an important consideration for side effects, may be higher than usual at equivalent intakes.

Between 500 and 600 cases of vitamin A adverse effects have been reported. The usual signs are peeling and redness of the skin, disturbed hair growth, loss of appetite, sickness and rarely liver injury. These have mainly occurred in patients with renal failure. Withdrawal of the vitamin A usually results in regression of the signs and symptoms in a matter of days, with no remaining effects after a few weeks. Hypercalcaemia has also been reported, but may be related to simultaneous intake of high levels of vitamin D. Whether hypervitaminosis A during pregnancy can give rise to fetal defects in humans is uncertain. However, in common with all other forms of medication caution should be exercised during early pregnancy.

The level of absorption depends on the vitamin A form (e.g. aqueous or oily). Since vitamin A is stored in the body, it is frequently *not given daily* for therapeutic purposes. Hence it is very difficult to define the highest daily level that can be regarded as safe. However, a maximum of about 150 times the RDA can be regarded as a reasonable indicator of the largest *single* oral dose of retinyl ester in oily form which would be tolerated without adverse effects. Although the evidence is inconclusive it would appear that any dose below about 15 times the RDA can be safely tolerated in long term administration.

β-Carotene, which is used extensively as a fat soluble food colourant, is a pro-vitamin A (page 112). In practical terms 6 μg β-carotene is converted to 1 μg of retinol. β-Carotene is poorly absorbed from the intestinal tract and conversion to vitamin A is retarded by feedback inhibition to such an extent that it is very difficult for manifestations of hypervitaminosis A to occur. Cases have occurred of carotene accumulation causing a reversible yellow colouration of the skin when excessively high doses of carotene containing foods have been consumed by food fads. It is said that carotenaemia may be associated with amenorrhoea but the evidence is not convincing.

The retinoids

For many years dermatologists have prescribed high doses of vitamin A for acne and some other rare skin conditions including Darier's disease and pityriasis rubra pilaris. These experimental treatments produced only partial success and, because the dose used was very high, they were often associated with toxic manifestations (page 119). In separate studies, it was found that retinoic acid, the metabolite of retinol, used topically could also produce interesting though variable improvement in some skin disorders. This led to the synthesis of various analogues of vitamin A of which two have now been the subject of extensive human experimentation, namely etretinate and isotretinoin (Figure 14).

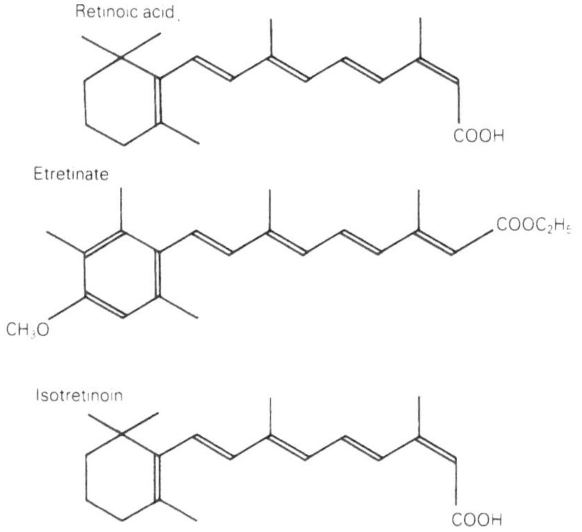

Figure 14   Formulae of retinoic acid and the retinoids (etretinate and isotretinoin)

Although these compounds are therapeutically effective they produce severe adverse reactions and have a low therapeutic ratio. IN CON-SEQUENCE THE PRESCRIPTION OF THESE TWO DRUGS IS CURRENTLY RESTRICTED TO SPECIALISTS AND SPECIAL PRECAUTIONS ARE ADVISED. IN SOME COUNTRIES USE IS RESTRICTED TO HOSPITALS ONLY. Neverthe-less, it is important that practitioners know something of their effects and toxicity.

Etretinate is used to treat patients with severe disorders of keratinization (often congenital in nature) as well as individuals with severe types of psoriasis. Treatment is started at a dose of 0.5 mg/kg twice a day, and then adjusted for the response of the individual patient. Continuous treatment is necessary to maintain the drug's effect. The main disorders of keratinization which appear to benefit most from etretinate are shown in Table 43. Patients suffering from these disorders are often very severely disabled both physically and socially, and the results of therapy can be extremely effective from both points of view despite the unpleasant side effects of the retinoid.

Table 43 Skin disorders for which etretinate is indicated
NOTE: ONLY FOR USE BY SPECIALIST DERMATOLOGISTS

Disorders of keratinization including:
    Darier's disease
    erythrokeraderma variabilis
    lamellar ichthyosis
    epidermolytic hyperkeratosis

All types of ichthyosis (if severe)

Psoriasis (if severe) including:
    pustular psoriasis
    erythrodermic psoriasis
    plaque-type psoriasis

Severe lichen planus

Multiple solar keratoses or basal-cell
    carcinomata

The toxic effects are mainly predictable (Table 44), dose-related, not life-threatening and reversible on discontinuation of treatment. Nearly all patients develop dry skins and mucosae, and about one third experience some loss of scalp hair, peeling of the skin of the palms and soles and softening of the nails. Plasma lipid levels may rise.

Pregnancy is an *absolute bar* to treatment and since the drug has a very long half-life, effective contraceptive measures must be taken both during therapy *and for a full year after stopping treatment with the drug.*

Isotretinoin may also be used to treat the disorders outlined for etretinate, but its major indication is cystic acne. Treatment is given on a continuous basis for about 4 months, at which stage there is a marked improvement in the clinical state in some 75% of those treated and the acne does not usually reappear even when therapy is stopped.

The side effects are similar to those with etretinate, but the teratogenicity is not so prolonged and contraceptive measures are not needed for so long after the drug is stopped.

Table 44   Side effects of the oral retinoids

Mucocutaneous
    dry skin
    facial or finger-tip dermatitis
    sensitivity to UV light (sunburn)
    poor wound healing
    cheilitis
    blepharitis
    mechanical fragility of skin and mucous
        membranes

Fetal
    teratogenicity

Musculoskeletal
    arthralgia
    premature closing of epiphyses
    diffuse spinal hyperostosis

Metabolic
    rise in fasting lipids (particularly in the
        obese and in women on oral contracep-
        tives
    minor increases in liver enzymes
    ? hepatic fibrosis

18

Vitamin D

Alternative names: calciferol; antirachitic vitamin

Vitamin D is the general term covering several related sterols which are present in various animal species. That present naturally in the human is vitamin D_3–cholecalciferol (Figure 15a), but ergocalciferol (vitamin D_2–Figure 15b) is often used therapeutically. This group of sterols is soluble in most organic solvents, but only poorly soluble in vegetable oils.

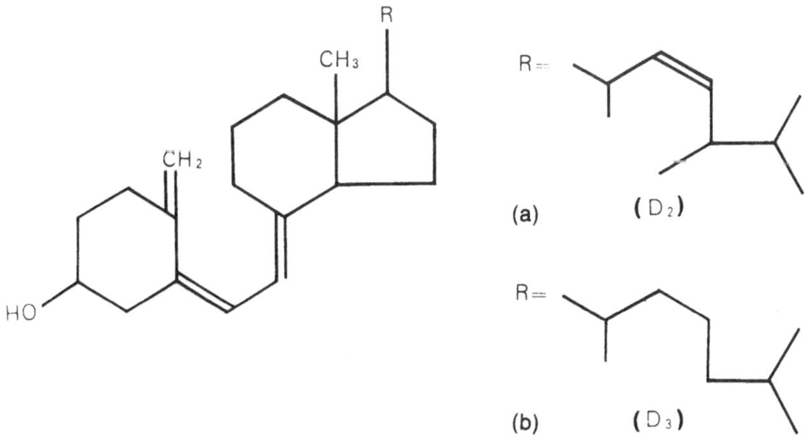

Figure 15 Structural formula of (a) cholecalciferol, (b) ergocalciferol

Sources

Vitamin D is not widely distributed in natural food sources, although high levels are found in the liver and viscera of fish and are present in dairy products. Due to supplementation the level is also high in margarine.

123

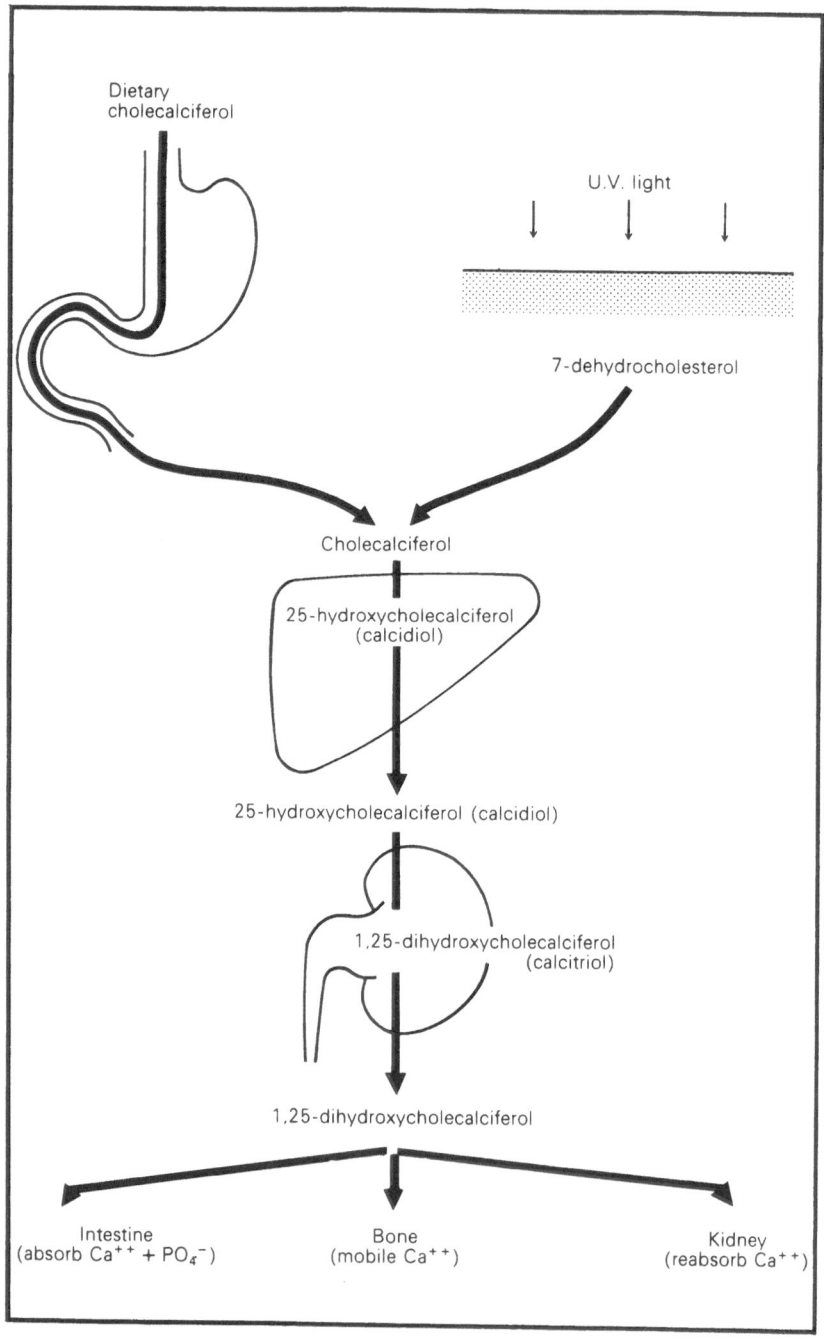

Figure 16 Formation and action of 1,25-dihydroxycholecalciferol

Vitamin D required by the body is mainly formed in the skin by the action of ultraviolet light on 7-dehydrocholesterol (Figure 16), and if there is adequate exposure of the skin to the sun dietary calciferol has relatively little importance.

Requirements

The normal human requirement is still unknown although it appears that it is of the order of 400 i.u. (10 μg) daily (Table 7). It is interesting to note that the authorities in many countries advise substantially lower values — probably in the expectation that the major share will be met by synthesis in the skin. Under normal conditions of exposure to sunlight, the precursor 7-dehydrocholesterol is converted into vitamin D_3 in the skin. When sun exposure is restricted dietary intake may be important. This is one of the very few vitamins where the levels should not be exceeded by a large factor due to the relatively narrow safety ratio (page 56).

A supplement of 5 μg is advised for both pregnant and lactating women.

Causes of deficiency

The inter-relationships which lead to the production of adequate levels of 1,25-dihydroxy vitamin D (calcitriol), the active hormone related to vitamin D, are shown in Figure 16. From this it is apparent that deficiency can occur as a result of inadequate exposure to sunlight if there are inadequate dietary levels. Recent studies have indicated that in countries where the weather is often inclement and UV exposure limited, the dietary component is vital. This particularly applies to immigrants from tropical countries and for those who for religious or other reasons cover a substantial portion of their body even during the period of sunlight (page 16). The conversion of either exogenous or endogenous calciferol to the active form involves appropriate mechanisms in the liver and kidney; the details of which are considered below.

Assessment of vitamin status

Vitamin D status can be determined best by the measurement of serum calcitriol, although at present most assessments involve the measurement of 25-hydroxy vitamin D (calcidiol). The vitamin D status can also be judged from the serum alkaline phosphatase level and from the bone appearance on radiography.

Activity

Although by tradition classed as a vitamin, vitamin D should now be regarded as a pro-hormone, particularly since, under normal circumstances most of the body needs can be formed in the body. Calciferols are absorbed from the intestinal tract in association with fats and this is aided by the presence of bile salts. However, of greater importance than the absorption of dietary calciferols, is the formation of vitamin D in the skin by the action of ultraviolet light on the precursor 7-dehydrocholesterol. Vitamin D is converted to 25-hydroxycholecalciferol (calcidiol) in the liver and then further hydroxylated to 1,25-dihydroxycholecalciferol (calcitriol) – the active hormone, often also termed 1,25-DHCC – in the kidney. The 1-hydroxylation in the kidney is under the control of the parathyroid hormone possibly through the intermediary of kidney phosphate levels (Figure 16).

1,25-Dihydroxycholecalciferol is concerned with the metabolism of calcium and phosphate via the formation of a calcium binding protein which assists the movement of calcium ions across membranes. The hormone increases the absorption of calcium and phosphate from the intestine; has a direct effect on calcification by increasing the uptake of minerals into bone, and within the kidney increases the clearance of phosphate.

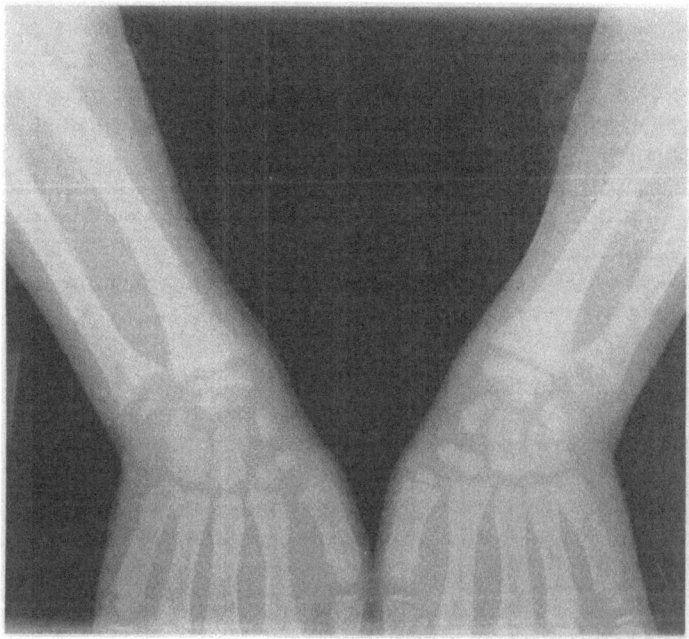

Figure 17 Radiograph of the wrists of a child with rickets. (By courtesy of Professor T. Sherwood, Addenbrooke's Hospital, Cambridge)

Clinical manifestations of deficiency

Deficiency of vitamin D in childhood leads to rickets, and in adults, when the epiphyses are fused, to osteomalacia. The two are differentiated solely by epiphyseal fusion. Hence in children the ends of the long bones show enlargements and deformities not seen in the adult (Figure 17).

Often the first symptoms to attract attention in rickets are excessive sweating and gastrointestinal disturbances. These are followed by signs of skeletal deformities: craniotabes (softening of the skull along the lamboidal sutures) often followed by 'bossing of the skull' (thickening of the cranial vault), when rickets occurs in the first few months of life (Figure 18). During

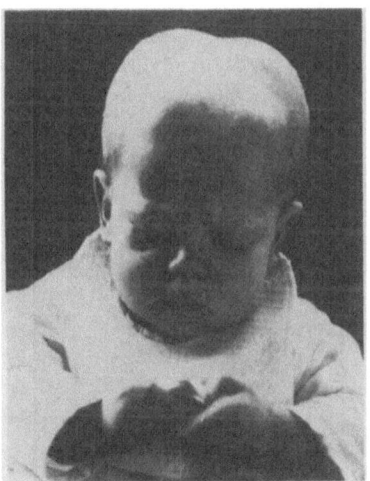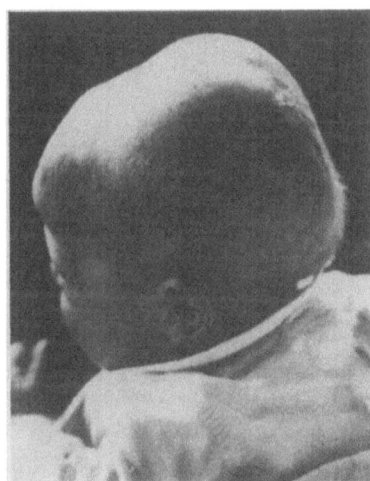

Figure 18   Vitamin D deficiency in male twins of 16 months showing typical 'skull bossing'

the next few months of life there is delayed primary dentition: rachitic rosary (enlargement of the costrochrondral junction) (Figure 19). Bow legs with a waddling gait, enlargement of the ends of the long bones, pelvic deformity and stunted growth are the manifestations seen in children over 2 years of age (Figure 20).

In osteomalacia there is gradually increasing bone rarefaction, particularly of the pelvis, leading to a typical severe deformity in which normal parturition is almost impossible, of the thorax, and of the cortex of the bones such that spontaneous fractures occur (Figure 21). In the elderly, an inadequate level of vitamin D is thought to be one factor in the pathogenesis of their fragile bones.

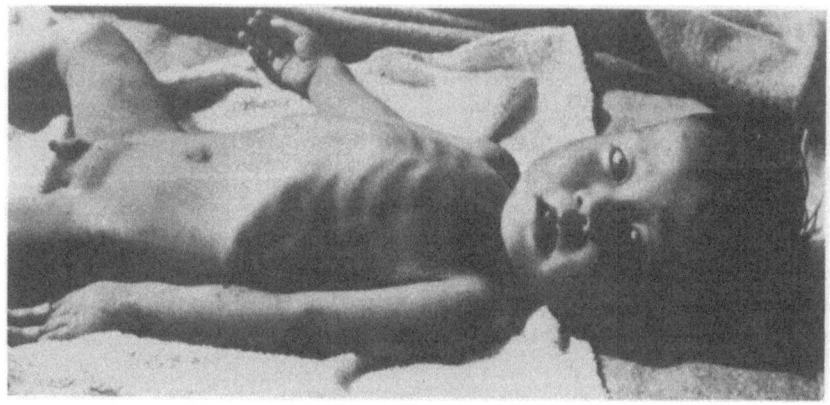

Figure 19 Rickety rosary in a young child. (Reproduced from 'Assessment of Nurtitional Status of a Community' by D.B. Jeliffe (1966) by permission of the World Health Organisation.

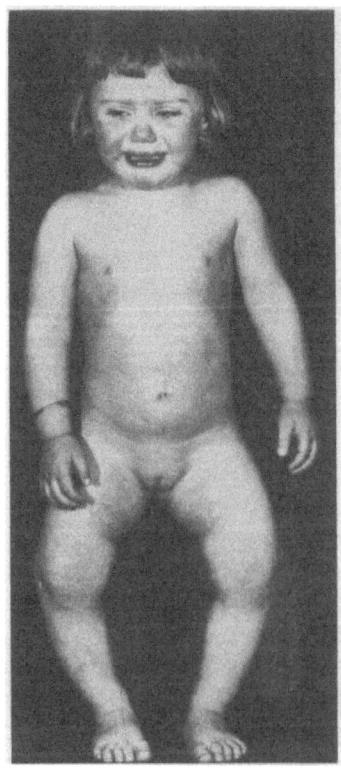

Figure 20 Pronounced bowing of the tibia in a rachitic child of 2 years. (By courtesy of Dr P. Hansell, Westminster Hospital, London)

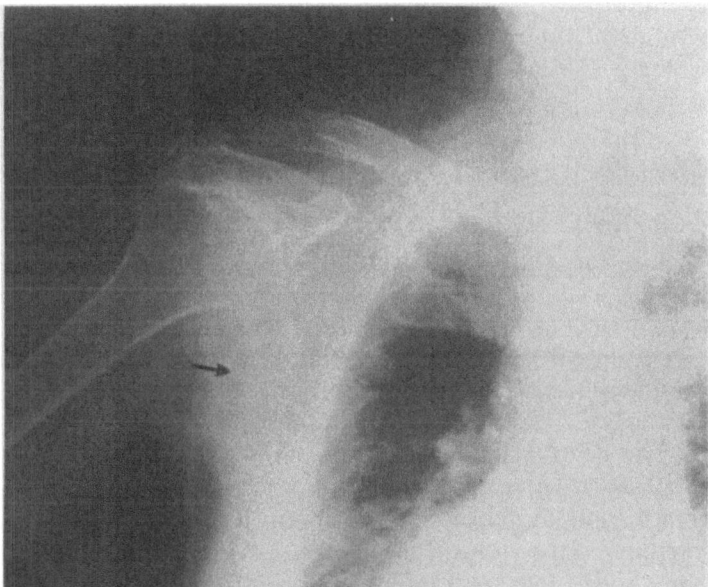

Figure 21 Radiograph of the shoulder in osteomalacia showing rarefaction of bones and spontaneous fracture of scapula (arrowed). (By courtesy of Professor T. Sherwood, Addenbrooke's Hospital, Cambridge)

In addition to the deformities of bone which stem from poor vitamin D production or intake, a large number of disorders (many of them rare) lead to failure of formation or action of calcitriol even when vitamin D is present

Table 45 Abnormalities of cholecalciferol metabolism which lead to hypocalcaemia

Disorders	Pathophysiology
Primary biliary cirrhosis Anticonvulsant therapy	liver abnormality leading to poor 25-hydroxylation
Chronic renal disease Cushing's disease Strontium poisoning Wilson's disease Cystinosis Fanconi syndrome	renal abnormalities leading to poor 1-hydroxylation
Some familial resistant rickets	absent 25-hydroxylase
Glucocorticoid therapy Familial hypophosphataemia	inhibition of 1-hydroxylase
Hypoparathyroidism	no parathyroid feedback
Certain malabsorption types of resistant rickets	reduced intestinal sensitivity to 1,25-dihydroxy cholecalciferol

(Table 45). Since these only respond to massive doses of vitamin D they have been termed 'vitamin D resistant rickets'. Nowadays these are treated with synthetic calciferol derivatives, particularly calcitriol (page 131).

Therapy

Deficiency states

The principal use of vitamin D is in the prophylaxis and treatment of the disorders of calcium–phosphorus metabolism. The two main disorders for which vitamin D is administered are rickets and osteomalacia. The prophylactic dose is about 500–1000 units daily, while for therapy doses of 5000 units daily may be given in infants, rising to 10 000 i.u. in severe cases and 15 000 in late onset rickets. Caution should be exercised in view of the relatively low safety margin for this vitamin.

Alternatively, massive dose prophylaxis and treatment is possible with a *single* oral or intramuscular dose of about 300 000 i.u. which becomes incorporated into the stores. This provides cover for about 6 months.

In infants and younger children spasmophilia may on occasion be due to a low blood calcium level and may be precipitated by vitamin D. To avoid this situation, the administration of vitamin D should be supplemented by simultaneous administration of oral calcium. In senile osteoporosis one of the factors is often a low intake of vitamin D, and there are merits in ensuring that all elderly people should receive adequate prophylactic amounts.

Other disorders

Hypoparathyroidism can be treated with high doses of vitamin D, which increases phosphaturia and concurrently plasma calcium. The dose should be judged by the plasma calcium levels and may require a maintenance dose in the range 50 000 – 200 000 i.u. per day.

Several syndromes have been described (page 129) in which bone rarefaction resistant to heroic doses of vitamin D is a feature (Table 45). These have been grouped together under the broad term 'vitamin D resistant rickets'. The majority of these result from inadequate or inappropriate metabolism of vitamin D into the hormone 1,25-dehydroxy-cholecalciferol (calcitriol). The use of modified forms of vitamin D for these indications is considered on page 131.

It has been recently suggested that in some patients cochlea deafness associated with otosclerosis is the result of chronic vitamin D deficiency and that the hearing of these patients can be improved with calciferol therapy.

Safety

Vitamin D is the vitamin most likely to produce adverse reactions, massive doses over a long period of time leading to calcification of soft tissues. Such deposits may be found in blood vessels, heart, lung or around joints; deposits in the kidney may lead to renal failure which has on occasion been fatal. In addition, general problems such as loss of appetite, weakness and constipation can occur.

Advice on vitamin D safety is difficult, not only because the extent of the synthesis in the skin is variable, but there appears to be a fairly wide range of susceptibility to the toxic action of vitamin D. Evidence of vitamin D intoxication has been reported with as little as 25 000 i.u. daily, although in most cases overt clinical signs of intoxication and hypercalcaemia occur when the dose exceeds 50 000 i.u. daily. Although the recommended dietary level for vitamin D is significantly below the 25 000 i.u. estimated to cause adverse reactions in some susceptible individuals, caution is advisable. This is particularly true in infants who are one of the main groups at risk. Other patients who are particularly sensitive to vitamin D are those with nephrocalcinosis and sarcoidosis.

Cholecalciferol related substances

Vitamin D can be regarded as a pro-hormone which is converted in the liver into the 25-hydroxy form (calcidiol) and subsequently hydroxylated one stage further in the kidney to form the active hormone 1,25-dihydroxy-cholecalciferol (calcitriol – page 126). Some of the disorders which reduce the extent of this conversion and reduce the therapeutic effectiveness of cholecalciferol or ergocalciferol are shown in Table 45 (page 129).

Recently several related substances, some hydroxylated and some in addition containing halogens, have been used experimentally in these rare disorders that do not readily respond to the vitamins themselves. Studies are still largely experimental and are best confined to specialist units.

Two preparations are currently available, alfacalcidol (1 α-(OH) D_3) and calcitriol. Both substances maintain plasma calcium in hypoparathyroidism, and are also effective in secondary hyperparathyroidism of chronic renal disease. They may also have a place in the treatment of hereditary hypophosphataemic rickets and vitamin D dependent rickets. Calcitriol currently seems to have advantages over alfacalcidol.

19

Vitamin E

Alternative name: α-tocopherol

The term vitamin E covers several related tocopherols which have been isolated from natural sources, but the most active of these is the α-form (Figure 22). It is an alcohol derived from phytol and trimethylhydroquinone, is soluble in organic solvents and in its non-esterified form is readily oxidized.

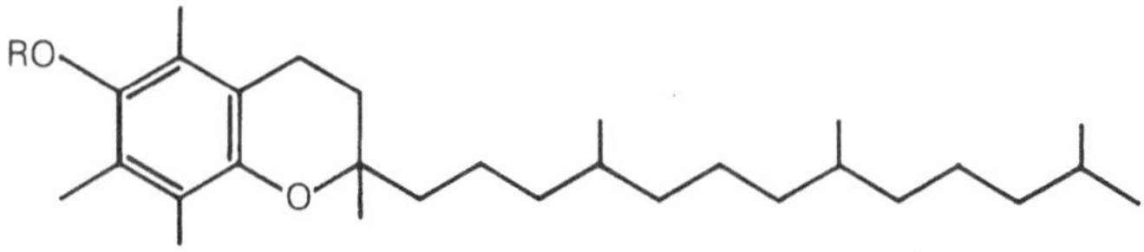

Figure 22 Structural formula of alpha tocopherol

Sources

Vitamin E is widely distributed in many plants but in low concentration. These tocopherols are present in higher concentrations in the embryos of many seeds.

Requirements

It was only relatively recently (1959) that α-tocopherol was recognized as essential for human nutrition. A daily recommended level has now been established in the USA and this is shown in Table 7. However, a recommended daily intake has still been established by relatively few national authorities. For adult males the advised level is 10 mg d-α-tocopherol daily. A small supplement is advised for pregnant and nursing women.

The requirement may in part depend on the lipids in the diet, particularly on the intake of polyunsaturated fatty acids, but there is no simple relationship between the two.

Causes of deficiency

The main causes of a deficiency of vitamin E (Figure 23) show considerable similarities to those for retinol. However, in contradistinction to vitamin A, vitamin E deficiency is not known to occur as a result of a dietary deficiency unless there are other predisposing factors (high intake of polyunsaturated fats). It appears that ascorbic acid both protects and helps to regenerate vitamin E so that vitamin E reduction may also be associated with low ascorbic acid values. One of the groups at greatest risk are premature infants.

Assessment of vitamin status

Vitamin E status can be assessed by the determination of serum tocopherol, from urinary creatine excretion and by the determination of red cell haemolysis by peroxide.

Activity

The absorption of vitamin E, like that of all fat soluble vitamins, is linked to fat absorption and is bile salt facilitated. Absorption across the placenta is poor and neonatal levels are low. In normal people about 70% of the intake is absorbed and in its turn it facilitates vitamin A and carotene absorption, but there is still doubt about its further metabolism. Vitamin E is stored in the liver and fatty tissue.

Vitamin E has a powerful antioxidant effect within the animal body, particularly for lipids. The important physiological actions may be stabilization of the mitochondrial membranes and maintenance of low cellular peroxide levels, particularly in the presence of high levels of polyunsaturated fatty acids. Its physiological effects are closely related to those of selenium, and roles in intracellular respiratory systems and nucleic acid metabolism have been postulated. However, despite intensive study, the true biochemical role has not yet been defined.

Clinical manifestations of deficiency

Although laboratory evidence of deficiency can be demonstrated in adults there are no physical signs that can be conclusively attributed to low vitamin E intake alone. Very rare cases of encephalomalacia due to vitamin E deficiency have been precipitated with parenteral nutrition high in polyunsaturated fats.

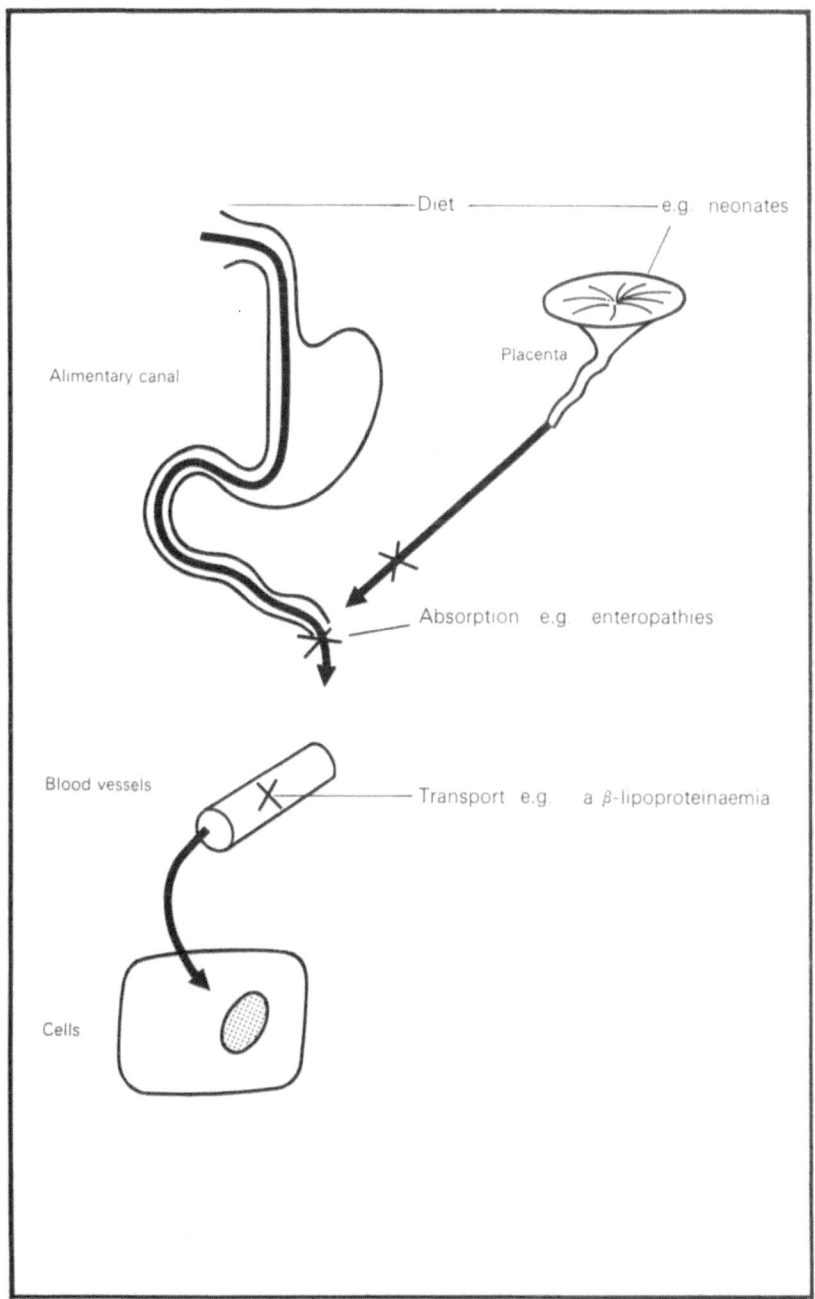

Figure 23   Mechanisms for hypovitaminosis

Similar neurological signs have been found in abetalipoproteinaemia and other disorders of fat absorption in which it is known that vitamin E levels are reduced (Table 46). It has, therefore, been suggested that central nervous system damage similar to that seen in animals can occur in such patients from the low vitamin E levels.

Table 46   The common neurological findings in human vitamin E deficiency

Areflexia
Cerebellar ataxia
Loss of vibration and postural sense
Pigmentary retinopathy
Ophthalmoplegia
Muscular weakness

Based on Muller *et al.* (1983)

In neonates, particularly prematures, a vitamin E responsive haemolytic anaemia can develop. There is also increasingly good evidence that the low levels of vitamin E are one factor in the genesis of retinal destruction in neonates exposed to high oxygen levels.

Therapy

Deficiency states

Supplementation is desirable in fat malabsorption or transport syndromes; in premature babies receiving artificial foods and when large amounts of polyunsaturated fatty acids are taken in the diet. The dosage varies from 5–20 mg daily in infants to about 30–100 mg in older children and up to 200 mg in adults on diets high in polyunsaturated fats.

Other disorders

Vitamin E has been used extensively in clinical medicine, but careful examination of all the available evidence reveals that in the majority of such disorders value has not yet been proven. In particular, there is no proven evidence of benefit in human muscular dystrophy despite earlier enthusiasm.

There is, however, well founded evidence for the value of large doses of vitamin E (400–600 mg per day over a prolonged period) in intermittent claudication. Among other disorders in which there is reasonable evidence for the administration of vitamin E (in doses up to 400 mg per day), even

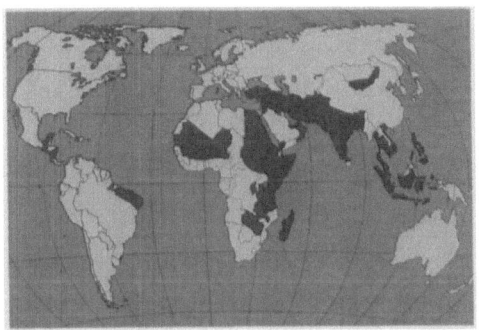

Plates 1–4 The main areas of prevalence of
vitamin deficiency diseases
Plate 1 Xerophthalmia

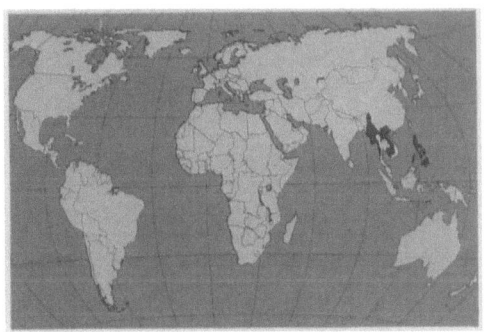

Plate 2 Beriberi

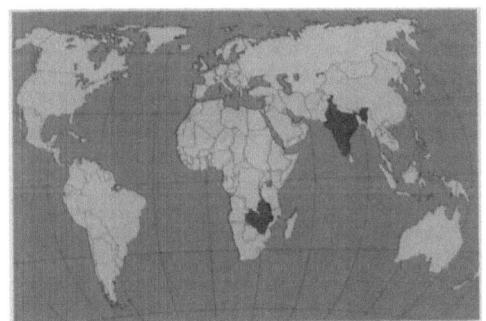

Plate 3 Pellagra

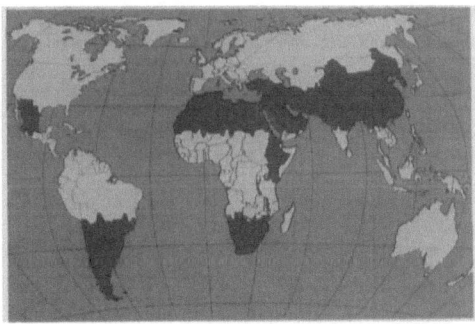

Plate 4 Rickets

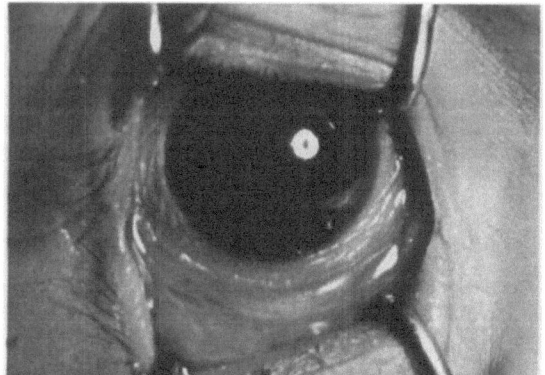

Plate 5 Xerosis. Note the hazy corrugated cornea, dry fatty conjunctival folds and Bitot's spot. Reproduced from *Assessment of Nutritional Status of a Community* by D.B. Jelliffe (1966) by permission of the World Health Organization

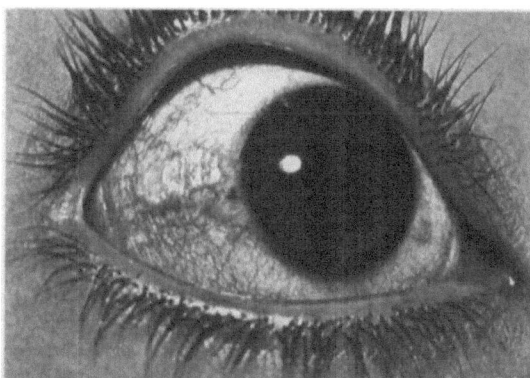

Plate 6 Example of typical Bitot's spots. Nine-year-old Jordanian male. Serum vitamin A 4 μg/100 ml ('normal' 20–50 μg) liver vitamin A 21 μg/g fresh liver ('normal' above 45 μg). Excellent response to large doses of vitamin A. Reproduced by permission of Professor D.S. McLaren, Nutrition Research Laboratory, American University of Beirut

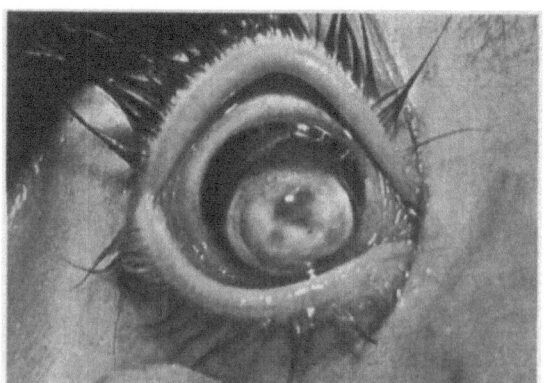

Plate 7 Keratomalacia. Fourteen-month-old Jordanian male child. Serum vitamin A μg/100 ml ('normal' 20–25 μg) liver vitamin A 3 μg/g fresh liver ('normal' above 45 μg). By courtesy of Professor D.S. McLaren, Nutrition Research Laboratory, American University of Beirut

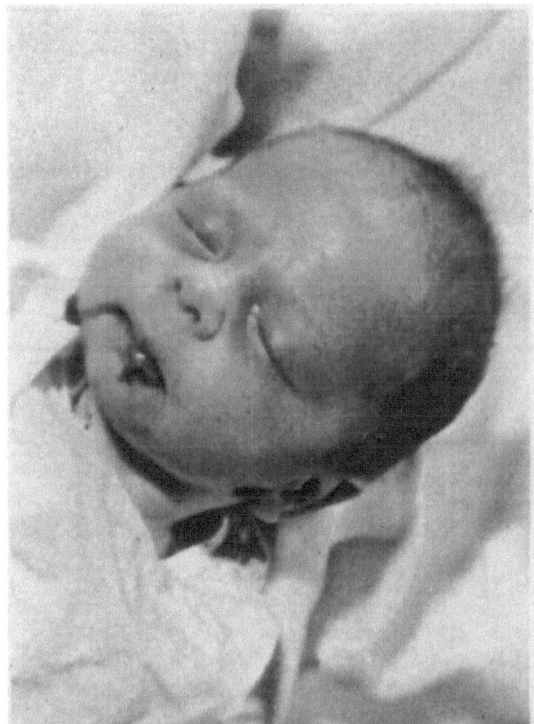

Plate 8 Blood-stained vomit in infant with haemorrhagic disease (vitamin K deficiency)

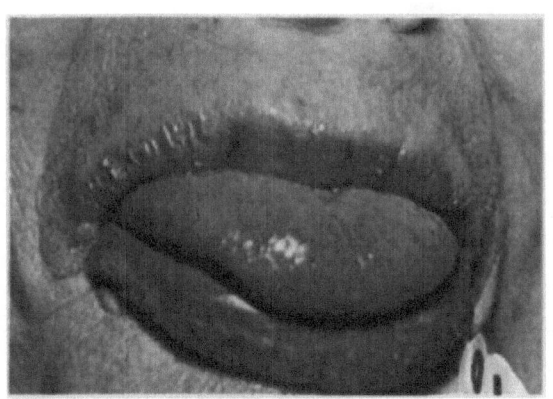

Plate 9 Riboflavine deficiency, showing angular stomatitis and severe cheilosis. Case seen in England 1966. Reproduced by courtesy of the Geoffrey Taylor Trust

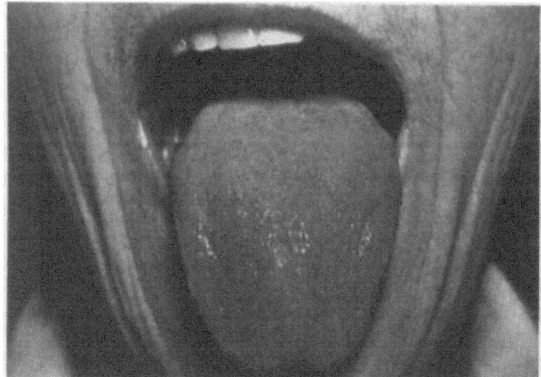

Plate 10 a. 'Normal'-tongue. b. So-called 'geographical tongue' which many believe represents early B group deficiency probably involving riboflavine deficiency as a main component. Note the patchy shedding of filiform papillae, some fissuring, and fungiform papillae at tip. c. Same patient as (b) after 3 months' treatment with a high dose B complex preparation ('Becosym Forte'). Reproduced by courtesy of the Geoffrey Taylor Trust

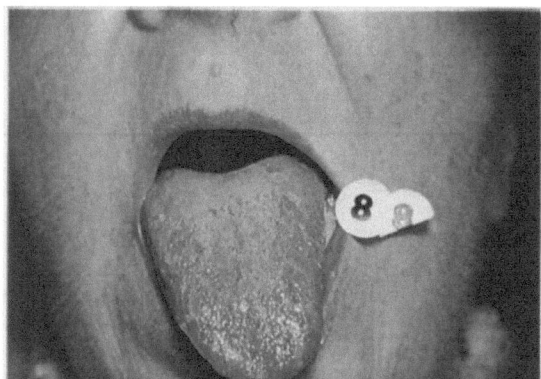

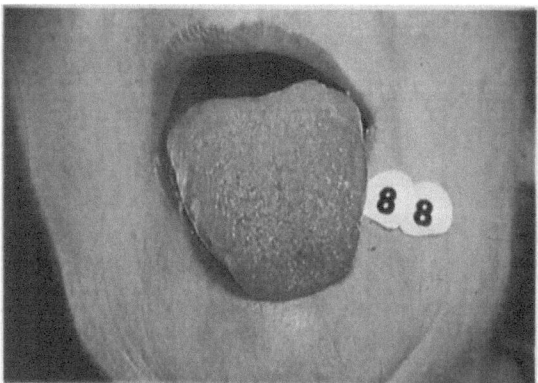

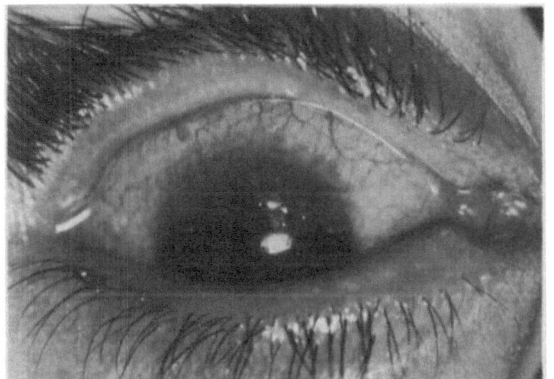

Plate 11 Vascularization of the cornea in riboflavine deficiency. Reproduced from *Assessment of Nutritional Status of a Community* by D.B. Jelliffe (1966) by permission of the World Health Organization

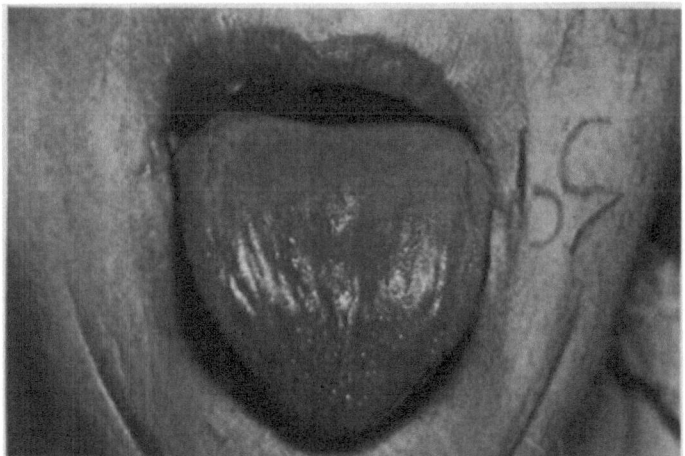

Plate 12 Typical tongue of nicotinic acid deficiency (compare with normal tongue Plate 10a). Note the red, raw, smooth appearance. Reproduced by courtesy of the Geoffrey Taylor Trust

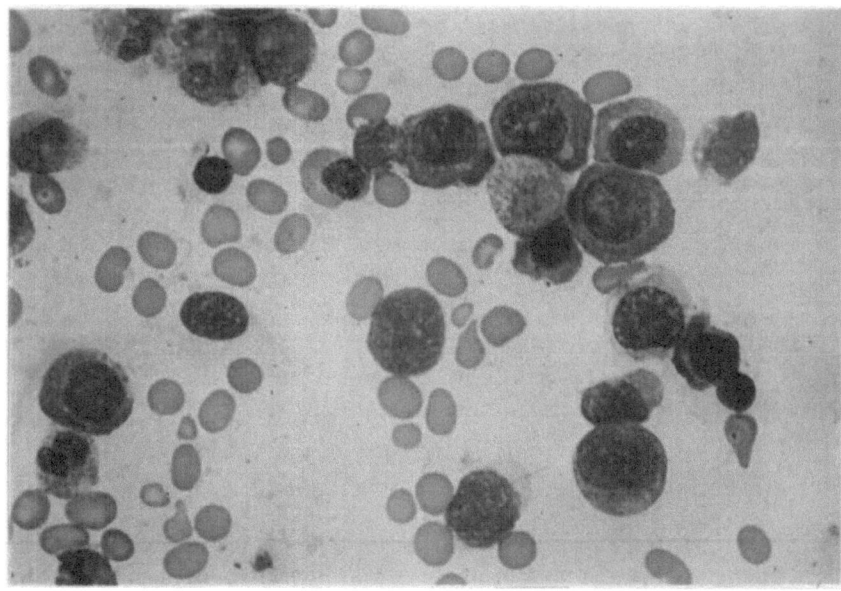

Plate 13 Megaloblastic bone marrow in vitamin B$_{12}$ deficiency, patient whose blood is shown in Plate 14. By courtesy of Prof. F.G. Hayhoe, Dept. of Medicine, University of Cambridge

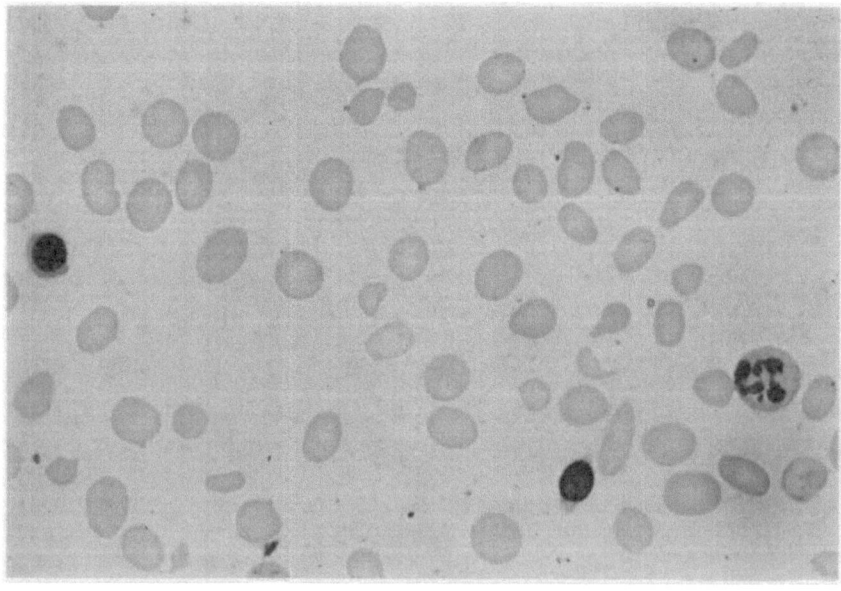

Plate 14 Typical blood picture in vitamin B$_{12}$ deficiency. By courtesy of Prof. F.G. Hayhoe, Dept. of Medicine, University of Cambridge

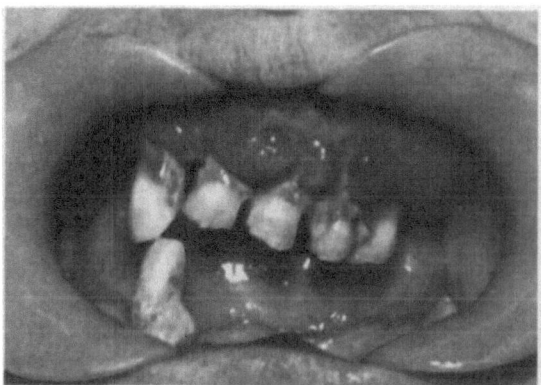

Plate 15 Swollen and bleeding gums in a middle-aged man with scurvy. Several of the teeth had become loose. By courtesy of Dr. Hansell, Westminster Hospital, London

Plate 16 The skin in scurvy showing hair follicles with unerupted curled hairs. Reproduced by courtesy of the Geoffrey Taylor Trust

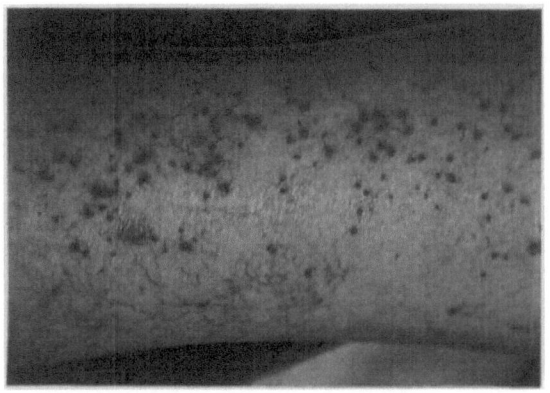

Plate 17 Skin petechiae in scurvy – photographed in England 1966. Reproduced by courtesy of the Geoffrey Taylor Trust

Plate 18 More extensive subcutaneous and intracutaneous haemorrhages in a case of scurvy. Reproduced by courtesy of the Geoffrey Taylor Trust

though therapeutic value has not been clearly proven, are certain fibrous tissue degenerations (e.g. Peyronie's disease, and Dupuytren's contracture).

There is evidence (though the reports are conflicting) that vitamin E (100 mg/kg daily orally) reduces the incidence of retrolental fibroplasia in premature babies who must receive oxygen for the relief of the respiratory distress syndrome. Vitamin E (25 mg/kg i.m.) may also be used for the prevention of intraventricular haemorrhage in prematures.

Safety

There is no evidence of adverse effects as measured by 14 different biochemical and clinical parameters, despite the fact that doses well over 100 times the RDA have been consumed for prolonged periods.

20

Vitamin K

Alternative name: phytomenadione, phylloquinone

A large |number| of naphthaquinones share vitamin K activity to different degrees, but the most active naturally occurring form is vitamin K_1 (Figure 24). The natural products are all fat soluble but water soluble derivatives showing most of the activity of the natural substances have been synthesized and used therapeutically.

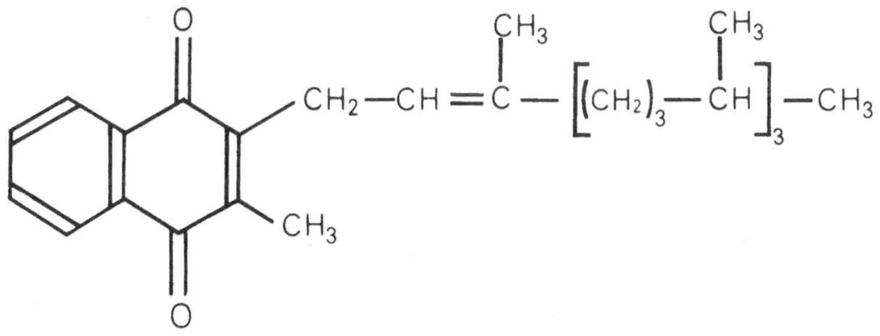

Figure 24   Structural formula of phylloquinone

Sources

Vitamin K is found widely in nature and particularly in brassicas and in animal liver. The main source of human vitamin K is believed to be synthesis by the intestinal bacteria. Transmission across the placenta is poor and the neonate has low levels.

Requirements

Any level in the dietary intake is probably superfluous because bacterial synthesis in the intestine provides about 1–1.5 mg vitamin K daily which appears to be the amount that covers the human requirements.

Due to a lack of adequate evidence, no national authority defines a specific RDA but the USA suggests that the range is 0.07–0.14 mg daily.

Causes of deficiency

Since vitamin K is synthesized in the intestine and the vitamin can then be absorbed, dietary sources of the vitamin are only necessary on very rare occasions. The main causes of a deficiency are shown in Table 47.

Table 47   The main causes of vitamin K deficiency

Reduced intake	– antibiotics
	– sterile gut of the newborn
Poor absorption	– obstructive jaundice
	– fat absorption defects
Poor utilization	– liver dysfunction, e.g. cirrhosis
	– antagonist activity, e.g. anticoagulants

Assessment of vitamin status

Vitamin K status is assessed by the determination of the specific clotting factors. It is also possible to measure serum phylloquinone.

Activity

Vitamin K is absorbed in association with the dietary fats, and absorption is facilitated by the presence of bile salts. Very little is stored and rapid depletion occurs with reduced intake.

The most important physiological function for vitamin K is the production of certain plasma coagulation factors in the liver: prothrombin; proconvertin (Factor VII); Christmas Factor (Factor IX); Stuart–Prower factor (Factor X). It probably facilitates the incorporation of sugars in proteins at a post-ribosomal stage. Several antagonists to vitamin K are now known, including related quinones, coumarin derivatives and indanediones. They are used therapeutically as oral anticoagulants, the effects being reversible with vitamin K_1 but not with the water soluble analogues.

Vitamin K also plays a significant role in calcification.

Clinical manifestations of deficiency

Low levels of vitamin K in the body can occur with any disorder of the liver or biliary system that reduces the amount of bile salts in the intestinal tract (Table 47). It also occurs when the absorptive capacity of the intestinal tract mucosa is reduced. Oral anticoagulants competitively inhibit the activity of vitamin K in the body and lead to functional low levels.

Low levels of vitamin K lead to a haemorrhagic diathesis. In neonates low vitamin K levels are common if no prophylactic dose is given, and small haemorrhages may occur subcutaneously. Occasionally they may occur in vital structures, even leading to neonatal deaths (See Plate 8).

Subcutaneous haemorrhages are also the most common manifestation in low vitamin K levels in adults. However, if the low levels are left untreated, extensive and fatal haemorrhages may occur. This occurs particularly when biliary tract abnormality is the cause of the low vitamin K level and surgical correction of the biliary tract abnormality is attempted without prior correction of the deficiency. Similar manifestations occur when excessive doses of oral anticoagulants are given.

There is also some evidence which suggests that osteoporosis and fractures of the neck or the femur may be related to low levels of circulating vitamin K. This requires further confirmation before vitamin K can be recommended for this indication.

Therapy

Deficiency states

In the neonatal period vitamin K_1 can be administered prophylactically post-partum for haemorrhagic diseases of the newborn at a dose of 0.5–1 mg. If necessary it can be given in the same dose therapeutically. It is preferable, however, to administer vitamin K_1 to the mother (5 mg i.m. or 10–20 mg orally) just before delivery to ensure that the baby's level is adequate.

Vitamin K_1 (1–5 mg i.m.) or synthetic analogues (10–40 mg by mouth daily) may be given when there is poor absorption of the vitamin from the intestinal tract. This is particularly important if surgery is contemplated, and should be regarded as the standard procedure prior to surgery of the biliary system.

When haemorrhage is due to low levels of the clotting factors formed under the action of vitamin K (page 140), 10 mg vitamin K_1 may be administered intramuscularly.

When anticoagulant therapy has lowered prothrombin to potentially dangerous levels or where a haemorrhagic diathesis has been produced by anticoagulant therapy, 5–10 mg vitamin K_1 may by given either by mouth, intramuscularly or by *slow* intravenous infusion. The prothrombin level should be estimated again about 3 hours later and a further dose given if necessary. When the level is reduced but not dangerously low, correction should be by low doses of vitamin K_1 if it is intended to resume anticoagulant therapy, since high doses will abolish the effect of subsequent doses of anticoagulants. It should be noted that water soluble synthetic analogues do not have a place in the modification of the results of anticoagulant therapy though they can be used for most of the other indications.

Safety

Vitamin K_1 is generally well tolerated. Side effects such as a sensation of heat in the head, dyspnoea and pain in the chest have occasionally been reported after contra indicated rapid intravenous injection, but are probably due to emulsifiers in the intravenous solution. Therefore, oral or intramuscular administration are preferred and very slow intravenous injections should be restricted to emergency situations.

High doses of menadione (vitamin K_3), and its water soluble derivatives have been implicated in producing haemolytic anaemia, hyperbilirubinaemia and kernicterus in the newborn with immature liver function, and erythrocyte haemolysis in patients with glucose-6-phosphate dehydrogenase deficiency; in such cases vitamin K_1 can be administered with complete safety.

21

Vitamin B$_1$

Alternative names: thiamine, aneurine, antineuritic vitamin

Vitamin B$_1$ consists of a pyrimidine ring and a thiazole linked by a methylene bridge and contains a quarternary nitrogen (Figure 25). It is water soluble, of good stability even to heat but sensitive to ultraviolet light.

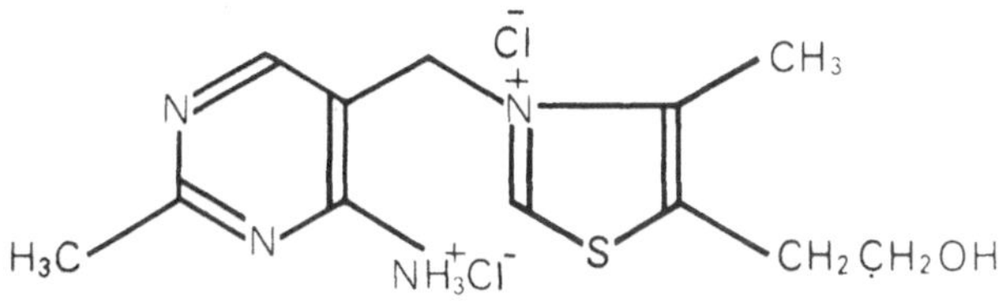

Figure 25 Structural formula of thiamine

Sources

Although intestinal bacterial synthesis of vitamin B$_1$ occurs in many animals, humans are probably almost entirely dependent on dietary sources. Vitamin B$_1$ is present in practically all plant and animal tissues and is present in high concentrations in yeast and in the pericarp and germ of cereals. In some countries, widely used cereal products are now enriched with vitamin B$_1$ and in these circumstances such sources probably form as much as 30–40% of the daily intake.

Requirements

The daily requirements for vitamin B$_1$ are shown in Table 7. The USA RDA for male adults is 1.4 mg and most countries advise a similar figure. The requirement is higher than normal when carbohydrate forms the major dietary component and less when fat and protein provide the large proportion of the daily calories. For this reason some tables of vitamin requirements are expressed in mg vitamin B_1 per 1000 carbohydrate kilocalories (USA, for example 0.5 mg per 1000 kcal or per 4200 kJ). The requirements are also increased during periods of increased metabolism (e.g. fever, hyperthyroidism, muscular activity) and a supplement of 0.4 and 0.5 mg, respectively are advised during pregnancy and lactation.

Causes of deficiency

Vitamin B_1 deficiency may arise either directly as a result of low intake of the vitamin or from disproportionate carbohydrate ingestion. During pregnancy increased tissue utilization may cause a deficiency which is often aggravated by loss of appetite and vomiting. Diseases that interfere with absorption of substances from the alimentary canal (e.g. sprue, idiopathic steatorrhoea, ulcerative colitis, etc.) may also produce a deficiency even when the dietary intake is apparently adequate.

A major cause of vitamin B_1 deficiency (and deficiency of other vitamins) in industrially developed communities is chronic alcoholism (page 71).

Tea (and to a lesser extent coffee) drinking can also cause vitamin B_1 deficiency (by the production of inactive vitamin B_1 disulphide), and in some parts of Thailand some $20-25\%$ of the population are deficient as a result of such dietary items as tea, betel nuts and raw fermented fish.

In Japan, on the other hand, not only is the vitamin B_1 level depressed as a result of thiaminase in the raw fish diet, but it has been found that in university students, thiamine deficiency is common due to the eating of convenience foods at a time when strenuous exercise increases the vitamin B_1 needs.

Assessment of vitamin status

Vitamin B_1 status is assessed by plasma or urinary thiamine or for the metabolic integrity by blood pyruvate levels or red cell transketolase.

Activity

Vitamin B_1 is rapidly and actively absorbed from the small intestine, and is phosphorylated into the active coenzyme – vitamin B_1 pyrophosphate (TPP; co-carboxylase). The body is incapable of storing the free vitamin, but small amounts of the phosphorylated form are present in all cells, including erythrocytes. Excess quantities are excreted in the urine as the free vitamin or a metabolite, and small quantities are also lost in the sweat. Vitamin B_1 pyrophosphate is an important coenzyme in mammalian tissues, and the reactions in which it plays a role are shown in Table 48. Hence vitamin B_1 is a vital factor in carbohydrate metabolism, and in vitamin B_1 deficiency blood pyruvate and often blood lactate levels rise steeply. It is not certain whether the central nervous system effects of vitamin B_1 deficiency should be attributed to these effects on carbohydrate metabolism or to the resulting decrease in active acetate for the production of acetyl choline. At very high doses vitamin B_1 is believed to exert an acetylcholine potentiating effect.

Table 48 Some important mammalian biochemical reactions involving thiamine pyrophosphates (TPP)

Reaction	Comments
Oxidative decarboxylation of pyruvic acid	also needs coenzyme A, lipoic acid and nicotinamide adenine dinucleotide (NAD)
Transketolation in pentose phosphate cycle	(1) provides pentose phosphate for nucleotide synthesis (2) supplies NADP for various synthetic pathways

Raw fish (via a heat labile thiaminase) and some bacteria (e.g. *Bacillus thiaminolyticus*) destroy B_1 and the raw fish diet is reputed to make about 3% of Japanese vitamin B_1 deficient. Direct antagonists which have been used in human experimental studies include oxythiamine and pyrithiamine.

Clinical manifestations of deficiency

In human experimental depletion studies, the first symptoms of a developing deficiency are depression, irritability, defective memory and the signs of peripheral neuritis. This stage is quite common in alcoholics. When deficiency develops naturally the prime symptoms and signs are related to the peripheral neuritis – paraesthesia, hyperaesthesia, muscle weakness and wasting. The risk of vitamin B_1 deficiency is increased by a high carbohydrate intake.

Gross vitamin B_1 deficiency is known as 'beri-beri' with three recognized types – 'dry' (Figure 26), cardiac oedematous – otherwise known as 'wet' (Figure 27) – and cerebral beri-beri (Wernicke's encephalopathy). The main manifestations of the three types are: neuritis; cardiomyopathy with cardiac enlargement and pronounced oedema; confusion, polyneuritis and sixth nerve paralysis (producing nystagmus). However, if an adequate clinical examination is made, most cases show some signs of the full pathology of vitamin B_1 deficiency. There are divergent views on whether Korsakoff's psychosis, usually seen with chronic alcoholism is a pure vitamin B_1 deficiency. Other factors are probably also involved.

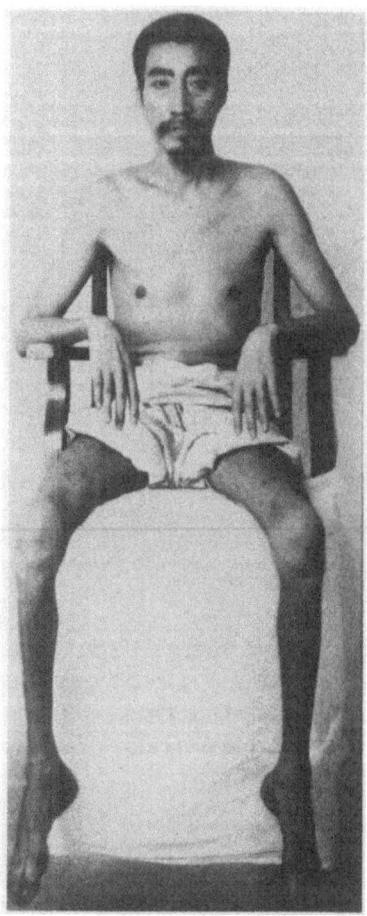

Figure 26 'Dry' beri-beri showing wrist-drop, foot-drop and marked wasting of lower extremities

A pure vitamin B₁ deficiency is rarely seen in clinical practice, most patients manifesting evidence of other deficiencies.

Therapy

Deficiency states

Vitamin B₁ is specific in the treatment of beri-beri and other disorders associated with vitamin B₁ deficiency. In mild deficiency states a daily dose of 100 mg is usually sufficient, but this can be increased up to 200 – 300 mg in severe cases.

146

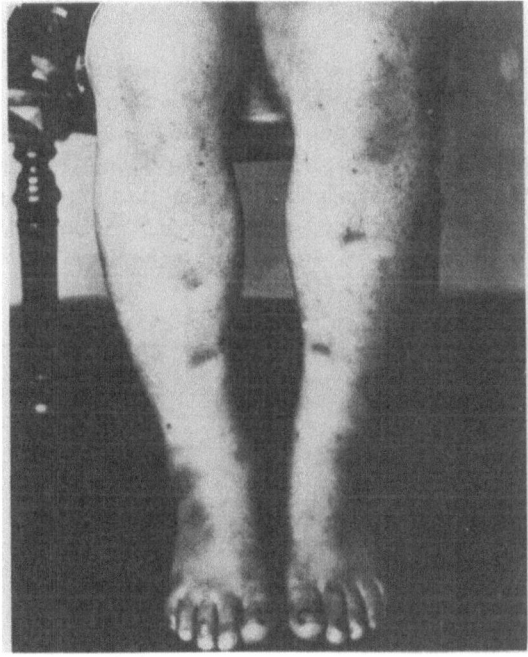

Figure 27 'Wet' beri-beri, showing extensive oedema of the legs

Since vitamin B₁ deficiency is frequently involved in the aetiology of peripheral neuritis, vitamin B₁ may be tried, sometimes with considerable benefit, in neuritis from other causes. Specifically neuritis accompanying alcoholism and pregnancy, in both of which conditions there is presumptive evidence of vitamin B₁ deficiency, often responds well to vitamin B₁. When alcoholism has led to delirium tremens, large doses of vitamin B₁ together with other vitamins, particularly those of the B complex, should be given by slow injection.

Other disorders

Although at present there is no rational explanation for its benefit, large doses of vitamin B₁ (100–600 mg daily) have been advised in the treatment of such diverse conditions as lumbago, sciatica, trigeminal neuralgia, facial paralysis and optic neuritis. The response is variable, but the author has seen this therapy used with success, presumably when the disorder is an unusual manifestation of a vitamin B₁ deficiency.

Safety

The only reaction found in humans is of the hypersensitivity type usually after injection of vitamin B_1. A skin test dose is advised before undertaking parenteral administration of vitamin B_1 in patients with a history of allergic reactions. The *parenteral* doses that have produced these reactions vary from 5–100 mg but are normally in the higher part of this range. Very rare transient hypersensitivity reactions have been reported after oral high doses (usually in the range of 5–10 g). Hence the safety factor is large (Table 21).

22

Vitamin B$_2$

Alternative names: riboflavine, lactoflavine

Vitamin B$_2$ is an isoalloxazine derivative with a ribitol side chain (Figure 28). It is an orange yellow crystalline compound, sparingly soluble in water, heat stable at normal temperatures, but unstable in alkali solution and on exposure to ultraviolet light.

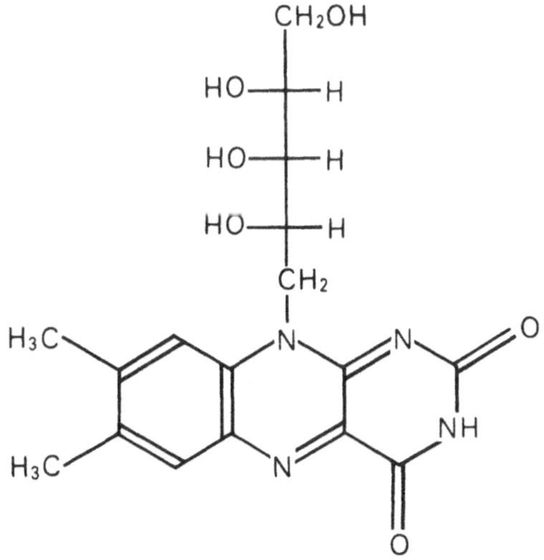

Figure 28   Structural formula of riboflavine

Sources

As the alternative name (lactoflavine) for this vitamin implies, vitamin B$_2$ is present in milk. It is also widely distributed in both animal and vegetable foods.

Requirements

Although vitamin B_2 can be synthesized by bacteria in the intestinal tract of many species, this does not form a significant source for humans and the daily requirements, shown in Table 7, must be met from the diet. Most countries define the adult male requirement in the range 1–2 mg (USA 0.6 mg/1000 kcal). Unlike thiamine there is no conclusive evidence of a link between the requirements and the intensity of metabolism. A supplement is advised for pregnant and nursing women.

Causes of deficiency

It is rare to find vitamin B_2 deficiency as an isolated and pure deficiency state. More commonly it is one deficiency in people who show general nutritional deficiencies. The causes are, therefore, those of any general deficiency state, as previously described.

Assessment of vitamin status

Riboflavine status can be measured by erythrocyte or urine riboflavine content or from the erythrocyte glutathione reductase level.

Activity

Vitamin B_2 is phosphorylated in the intestinal mucosa during absorption. It is stored in small quantities in liver, spleen, kidney and heart muscle. It is eliminated in the urine, and small quantities are lost in the sweat.

Table 49 Some important mammalian biochemical reactions involving flavoproteins derived from vitamin B_2

Reactions	Comments

Reactions in electron transport chain:
  (1) link between pyridine nucleotide systems and cytochromes
  (2) link between intermediary metabolites and cytochrome system
	both FAD and FMN are involved
	integral role in the biological oxidation system

Direct oxidase flavoprotein:
  (1) hypoxanthine – uric acid
  (2) aldehydes (e.g. retinal) – acid
  (3) amino acid oxidases, e.g. amino acids – ammonia + keto acids
	uses xanthine oxidase

Vitamin B$_2$ forms two tissue coenzymes, flavine mononucleotide (FMN) and more commonly flavine adenine dinucleotide (FAD). These in turn form the prosthetic groups of several different enzyme systems (the so-called flavoproteins) concerned with hydrogen transport (oxidation) (Table 49) particularly in the metabolism of amino acids.

Whereas vitamin B$_2$ is present in most animal tissues as the phosphate or one of the coenzymes it is present in the retina as free vitamin B$_2$. The significance of this is unknown.

Although vitamin B$_2$ is essential for growth and life, the human signs of mild degrees of deficiency are not impressive. This is surprising in view of its widespread metabolic functions. Antiriboflavine activity is found if the ribityl group is changed to another sugar (e.g. galactityl) or if the pyrimidine ring is substituted by 2:4 dinitrobenzene.

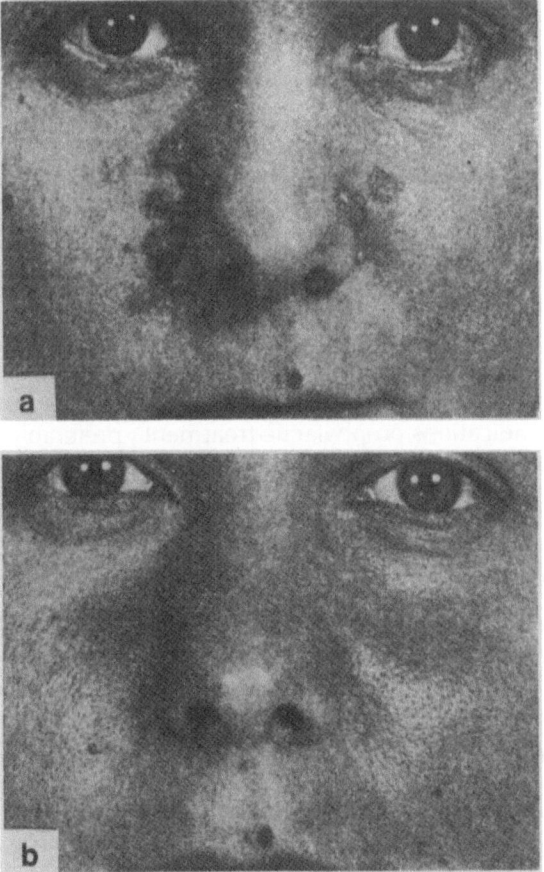

Figure 29 (a) Recurrent bilateral eczema of the nose and nasolabial fold; (b) response to riboflavin

151

Clinical manifestations of deficiency

The typical early signs of vitamin B_2 deficiency are related to oral and ocular lesions. The mouth shows cheilosis (Plate 9) (dryness, pallor, cracking of the lips) and angular stomatitis (transverse raw fissures at the mouth angles) and tongue denudation (perhaps patchy as 'geographical tongue') (Plate 10). Eye symptoms at this stage include photophobia, lachrymation and a sensation of 'grittiness' under the lid.

In later stages of deficiency there is a seborrhoeic dermatitis often maximal around the nasolabial fold (Figure 29) and a vulval or scrotal dermatitis. The eye shows a bilateral scleral vascularization (Plate 11) and corneal opacities may develop. Bone marrow aplasia with resulting anaemia has also been reported.

Therapeutic uses

Deficiency states

Vitamin B_2 in daily oral doses of 10 mg is employed for the treatment of the clinical signs and symptoms of vitamin B_2 deficiency described above. Since it is usual to find a multiple vitamin deficiency in those patients who show evidence of vitamin B_2 deficiency, a preparation containing other members of the B complex usually has advantages. Recovery from vitamin B_2 deficiency tends to be rather slow.

These lesions, which are suggestive of vitamin B_2 deficiency, are not uncommon in patients who have undergone gastrectomy, and have also been reported in patients under treatment with broad-spectrum antibiotics. In both these indications prophylactic treatment , preferably with a mixed vitamin B complex preparation containing a dose of 5 mg vitamin B_2 daily, is probably wise.

Other disorders

Surgery or trauma to the eye may cause corneal vascularization and therefore vitamin B_2 is often given prior to ocular surgery, particularly if the patient shows any signs of malnutrition or, for example, in the elderly.

Vitamin B_2 (25 mg per day) has been reported to be effective in cramps, though the rationale is not apparent. Vitamin B_2 has also been used successfully (100 mg daily orally) in the muscle weakness associated with the rare cases of NADH dependent respiratory enzyme deficiency.

Claims have been made that vitamin B_2 is of value in the treatment of trachomatous panus, angular blepharitis and phlytenular keratitis, but these do not stand critical analysis. Claims have also been made, though without any rationale, that vitamin B_2 may sometimes be effective in the treatment of migraine. In any of these disorders where there is no logical explanation for possible benefit, the usual dose to be tried is 30 mg daily.

Safety

No instances of adverse reactions with this vitamin have been reported despite high dosages.

23

Vitamin B₆

Alternative name: pyridoxine

The term vitamin B_6 is used to cover a group of compounds that are metabolically interchangeable, pyridoxol (the alcohol), its aldehyde (pyridoxal) and its amine (pyridoxamine) (Figure 30). They are all colourless crystals soluble in water and alcohol, resistant to normal heat but decomposed by alkalis and ultraviolet light.

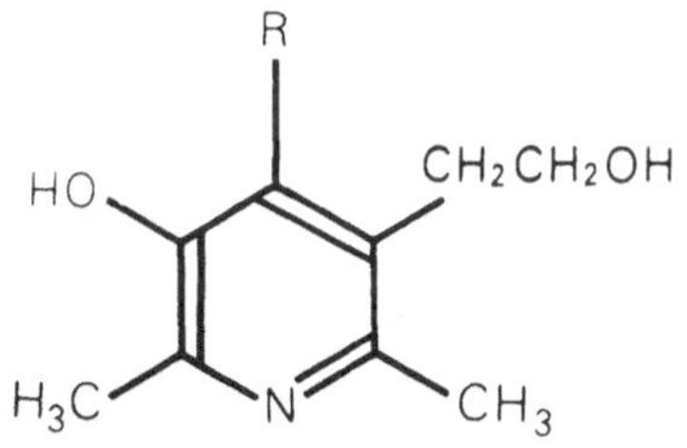

Figure 30 Structural formula of pyridoxine: pyridoxol, $R = CH_2OH$; pyridoxal, $R = CHO$; pyridoxamine, $R = CH_2NH_2$

Sources

The three forms are all widely distributed in low concentrations in all animal and plant tissues.

Requirements

Vitamin B_6 can be synthesized by bacteria in the intestinal tract but this source supplies negligible amounts to the body economy, and the requirements (Table 7) must be met from the diet. For adult males the USA RDA is 2.2 mg daily and a similar figure is advised by other national authorities. The requirement is increased when there is a substantial intake of protein in the diet (USA: 0.02 mg/g protein). Rather high additions to the RDA (0.6 and 0.5 mg, respectively) are recommended for pregnant and nursing women.

Causes of deficiency

Due to the wide distribution of vitamin B_6 in plant and animal foodstuffs, a true deficiency state due to an inadequate diet is rarely seen. An apparent deficiency may, however, occur in such conditions as the vitamin B_6 dependency states (where the requirement is high), in pregnancy and during the use of contraceptive pills. It may also be precipitated by certain drugs (e.g. isoniazid).

Table 50 Some important mammalian biochemical reactions involving pyridoxine coenzymes

Reaction	Comments
Transamination	allows synthesis of some amino acids from carbohydrate intermediates. Essential role in urea formation
Codecarboxylation	most amino acid decarboxylases are pyridoxal phosphate dependent. They form the important brain amines
Kynurinase activity	necessary for formation of nicotinic acid from tryptophan
Activity of certain deaminases	for serine, threonine, cystathionine, homoserine
Desulphydrases – transulphurase activity	for the interconversion and metabolism of sulphur containing amino acids
Linoleic acid → arachidonic acid	metabolism of essential fatty acids
Biosynthesis of coenzyme A	essential for interconnections involving 'active acetate'
Glycogen phosphorylase activity	glycogen → glucose-1-phosphate
Threonine aldolase activity	
Succinyl CoA + glycine → α-amino laevulinic acid	biosynthesis of porphyrin
Transmethylation of methionine	most methyl transfer reactions
Incorporation of iron in haemoglobin	

Assessment of vitamin status

There are several methods for determining the vitamin B$_6$ status. These include determination of the blood pyridoxal phosphate, urinary excretion of 4-pyridoxic acid, transaminase activity or the tryptophan load test.

Activity

Vitamin B$_6$ is rapidly absorbed from the small intestine and distributed throughout the tissues as the coenzyme. Excretion occurs in the urine mainly as the metabolite, 4-pyridoxic acid.

Vitamin B$_6$ in any form is rapidly converted in the body into the coenzymes pyridoxal-5'-phosphate and pyridoxamine phosphate. They form the prosthetic groups of enzyme – coenzyme systems in both protein metabolism and general metabolic processes (Table 50). As a result of these reactions, vitamin B$_6$ coenzymes play an essential role in energy production (supplying metabolites to the Kreb's cycle), protein and fat metabolism.

Vitamin B$_6$, as the coenzyme for decarboxylation of amino acids, has an important role to play in brain metabolism. In particular it is required for the formation of the whole group of amines that act as synaptic transmitters in various regions of the brain (noradrenaline, adrenaline, tyramine, dopamine and 5-hydroxytryptamine) and for the inhibitory γ-amino butyric acid.

Antipyridoxine activity is found in desoxypyridoxine, a tuberculostatic drug (isoniazid), an antihypertensive (hydrallazine), an antibiotic (cycloserine) and penicillamine (used in the treatment of the rare Wilson's disease).

Clinical manifestations of deficiency

Clinical signs of vitamin B$_6$ deficiency are rare even when there is clear biochemical evidence of deficiency. The only clinical manifestations attributable to vitamin B$_6$ deficiency are: convulsions in infants (probably due to low levels of γ-amino butyric acid); peripheral neuritis (on isoniazid therapy); and rarely a vitamin B$_6$ responsive anaemia; urticaria (in familial xanthurenic acid disease) and mental retardation (in familial cystathionin-uria).

Therapy uses

Deficiency states

It is rare to find a vitamin B$_6$ deficiency due to an inadequate diet. Treatment with isoniazid and related compounds can lead to a vitamin B$_6$ deficiency and vitamin B$_6$ dependency states also respond to high doses of this vitamin. In such patients the dose should be 40–200 mg per day.

Some of the manifestations of deficiency of other members of the B complex, e.g. cheilosis, hypochromic anaemia, neuritis, are found to respond better if a mixed B complex therapy is given (including vitamin B_6) than if the deficiency is treated with an individual member of the group.

Vitamin B_6 is required in increased amounts during pregnancy and while taking steroid contraceptive pills (25 – 50 mg daily), and it has also been used with benefit in the treatment of hyperemesis gravidarum in doses of 100 – 200 mg per day.

Other disorders

Vitamin B_6 has been advised in doses up to 300 mg per day, for the prevention and treatment of nausea and vomiting due to irradiation, drug therapy, anaesthesia and in travel sickness, but the response is variable.

Vitamin B_6 has also been used with benefit in patients with premenstrual tension, particularly those associated with depressive manifestations. Treatment should be started 3 days before the expected onset of the period at a dose of 50 – 75 mg twice a day. It should be continued until 2 – 3 days after the start of the period. The depression associated with coeliac disease is also reported to respond to vitamin B_6 80 mg daily.

Recently it has been suggested that the carpal tunnel syndrome may be related to chronic vitamin B_6 deficiency, and though not yet adequately confirmed, there is some evidence that the disorder can be treated with vitamin B_6.

Vitamin B_6 has also been advised for the relief of Chinese restaurant syndrome (due to high intake of monosodium glutamate). This also needs confirmation.

Safety

Daily oral doses of vitamin B_6 up to 50 times the RDA for periods up to 3 – 4 years and even higher doses for shorter periods have been administered without adverse reactions. Rare transient dependency states, consisting of ill-defined symptoms including nervousness and tremulousness have been induced when about 100 times the adult RDA was given for several weeks and then suddenly withdrawn. A recent report describes a sensory neuropathy at dramatically high vitamin B_6 dosage (1000 – 3000 times the RDA).

24

Niacin

Alternative names: nicotinic acid and amide, pellagra preventive (PP) factor, antipellagra vitamin, vitamin B_3

Niacin is pyridine carboxylic acid and nicotinamide is the corresponding amide (Figure 31). Both are water soluble, white crystalline substances, heat stable but labile to air, alkali or light.

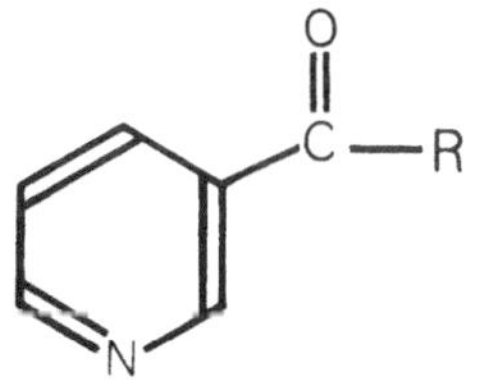

Figure 31 Structural formula of nicotinic acid, R = OH; nicotinamide, R = NH$_2$

Sources

Niacin is present in most animal and vegetable foodstuffs, but is particularly rich in meat, fish and wholemeal wheat flour. However, it is frequently present in food in a form from which it is not easily absorbed in the intestinal tract (e.g. maize, but from which it can be released by, for example, lime water as in the making of tortillas in Mexico).

Niacin biosynthesis from tryptophan takes place in the human if adequate essential amino acids, together with thiamine, pyridoxine and biotin, are present in the diet. Hence the availability of niacin depends in part on the levels of other vitamins in the diet.

159

Requirements

A small amount of the daily requirement of niacin is synthesized by intestinal bacteria but the majority is either taken in as niacin itself from the diet or synthesized in the body from L-tryptophan, an amino acid. This synthesis requires pyridoxal-5'-phosphate dependent enzymes and hence adequate dietary intake of pyridoxine. 60 mg L-tryptophan yields 1 mg niacin, and the dietary tryptophan normally accounts for about two thirds of the daily requirements which are shown in Table 7. Individual advice from national authorities differs little from this figure. Supplements are advised for pregnant and nursing women.

Causes of deficiency

Niacin deficiency may be one of the features of a general nutritional deficiency state, but is rare as an isolated deficiency. However, the exception to this is in maize-eating areas where pure niacin deficiency may occur even when the diet appears to be adequate.

Assessment of vitamin status

The only current method for assessing niacin status is by the determination of its metabolites, N'-methyl-nicotinic acid amide and 1-methyl-5 carbox-ylamide-2-pyridone in the urine.

Activity

Niacin is readily absorbed from the intestinal tract, and forms coenzymes with no true stores. Excretion of a little unchanged niacin occurs in the urine but the main metabolite is N-methyl nicotinamide.

Within the body niacin is converted to the nicotinamide enzymes:

> Nicotinamide adenine dinucleotide (NAD) or Coenzyme I (previously called diphosphopyridine nucleotide – DPN), and Nicotinamide adenine dinucleotide phosphate (NADP) or Coenzyme II (previously called triphosphopyridine nucleotide – TPN).

These are coenzymes concerned with the transfer of hydrogen by the hydrogenases and as such play a vital role in intermediary metabolism. A list of the more important dehydrogenases using NAD or NADP as coenzymes is given in Table 51 and in addition to these, NAD is the coenzyme for UDP-D-glucose 4^1-epimerase by which galactose 1-P is converted to glucose 1-P and hence to glycogen.

Niacin also has two pharmacological actions at high dosage: peripheral vasodilation (mainly as nicotinic acid) and serum cholesterol reduction.

Table 51 Dehydrogenases requiring:

NAD	NADP
Soluble α-glycerophosphate	glucose-6-phosphate
Lactic (and hence in	D-iso-citric (in mitochondria)
pyruvate metabolism	malic decarboxylase
Malic	SH-glutathione
D-Glyceraldehyde	NADH
phosphate	FADH
β-Hydroxybutyric	
Glucose	
L-Glutamic	
β-Hydroxy fatty-acyl-CoA	
NADPH	
FADH	

3-Acetyl-pyridine can act as an antagonist, as can the tuberculostatic – isoniazid. Leucine found in high concentrations in millet, one of the staple foods in India, increased the requirement for niacin

Clinical manifestations of deficiency

Niacin deficiency leads to the florid manifestations of pellagra, often summarized in the mnemonic 'Diarrhoea, Dermatitis and Dementia'. The gastrointestinal symptoms are often the first to appear and include glossitis and stomatitis, the tongue having a characteristic swollen and 'beefy-red' appearance (Plate 12), with anorexia, abdominal discomfort and diarrhoea. The pellagrous dermatitis is typically symmetrical and confined to the parts of the body that are exposed to sunlight; initially dry, discoloured and scaly but later desquamating (Figure 32). Early mental symptoms include lassitude, depression and loss of memory, but it may develop to a confusional state resembling Wernicke's encephalophathy. In rare cases the mental disorder may occur without the other signs – the so-called 'pellagra sine pellagra'. It is rare to see a pure niacin deficiency in clinical practice, since a deficiency of other members of the B complex usually coexists.

Hartnup disease is a rare familial disorder in which altered amino acid metabolism leads *inter alia* to niacin deficiency and clinical features resembling pellagra.

Therapeutic uses

Deficiency states

Nowadays niacin is almost invariably administered as the amide to avoid

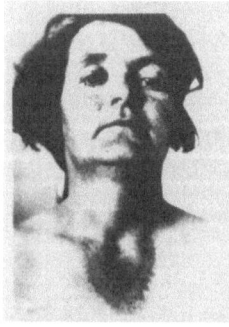
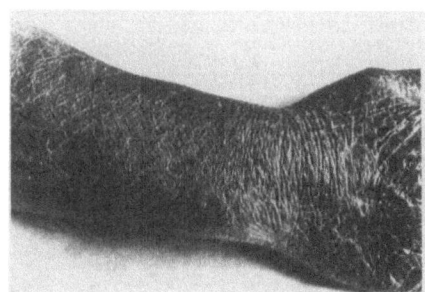
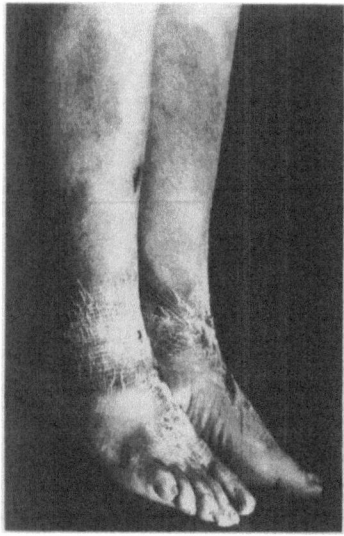

Figure 32 Pellagra dermatitis seen in this country in an elderly woman. Note characteristic changes in hands and feet and Casal's necklace

the undesirable skin flushing, and doses of 50 mg daily can be given when niacin deficiency is suspected, e.g. various disorders of the gastrointestinal tract, including sprue. When pellagra signs and symptoms are manifest, nicotinamide is given at a dose of up to 500 mg per day. In such patients it is usual to find a multiple deficiency of several members of the B complex, and in consequence the best results are usually found when pellagra is treated with a mixed B complex preparation which contains niacin, thiamine and riboflavine in high dosage. This is particularly true when peripheral neuritis and ocular signs are prominent.

Other disorders

Niacin (50 to 100 mg daily) has been administered with benefit in Hartnup disease. The vasodilator properties of nicotinic acid may be employed in the treatment of various conditions in which vasospasm is part of the disorder, e.g. peripheral vascular diseases, migraine. The dose is normally 100–300 mg per day. High doses of niacin (or the amide) of the order of 1–3 g per day have been used successfully for cholesterol reduction, though the side effects reduce the usefulness of this form of therapy. Similar high dosage has been advocated for the treatment of acute schizophrenia, but with modern neuroleptic drugs nicotinamide is not now used.

Safety

A transient tingling or flushing sensation in the skin after relatively large doses of nicotinic acid is a rather common phenomenon. The related substance nicotinamide only very rarely produces this reaction.

Doses of 200 mg to 10 g of nicotinamide daily have been used therapeutically under medical control for periods up to 10 years or more. In isolated cases, liver disorders, rashes, dry skin, liver damage and excessive pigmentation have been seen. The tolerance of glucose has been reduced in diabetics and patients with peptic ulcers have experienced increased pain. No serious reactions have been reported, however, even with these very high doses of nicotinamide.

25

Vitamin B₁₂

Alternative names: cobalamin, cyanocobalamin, extrinsic factor of Castle, animal protein factor

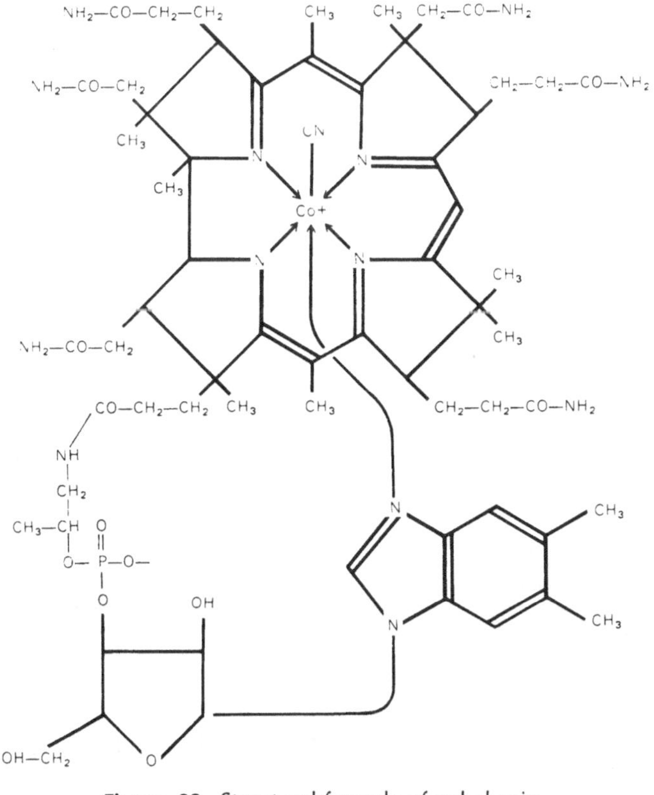

Figure 33 Structural formula of cobalamin

Vitamin B_{12} is the term applied to a group of chemically related metabolically equiactive compounds with a central 'corrin' ring system (porphyrin-like) surrounding a cobalt atom. The structure of the principle compounds with mammalian activity is shown in Figure 33. In natural form, vitamin B_{12} is probably bound to protein.

They are red crystalline hygroscopic substances, freely soluble in water and alcohol and labile in strong acid, alkali and light. They are stable alone at room temperature, but decompose in the presence of ascorbic acid, thiamine and nicotinamide so that it is difficult to prepare vitamin B_{12} containing multivitamin products.

Sources

With the exception of the liver, amounts in food are small and the vitamin is almost entirely absent from vegetables.

Requirements

National recommended daily intake for vitamin B_{12} is of the order of $3-5$ μg (Table 7). There is some synthesis by intestinal bacteria but absorption from any source is poor.

Causes of deficiency

Vitamin B_{12} deficiency can occur as a result of the conditions shown in Table 52. The most common cause is pernicious anaemia, in which there is a deficiency of the mucoprotein 'intrinsic factor' which protects vitamin B_{12} during its passage down the alimentary canal, therefore, facilitating absorption.

Table 52　The main causes of vitamin B_{12} deficiency

Reduced content in food	:	rarely seen
Reduced absorption	:	intrinsic factor deficiency, some worm infestations
Reduced storage and utilization	:	liver disorders

Assessment of vitamin status

While vitamin B_{12} status can be measured from the plasma level or an absorption test, it is usually assessed clinically on the basis of the macrocytic anaemia, hypersegmented leukocytes, megaloblastic marrow and achlorhydria. Other metabolic methods of assessment include the deoxyuridine test and methyl malonic acid in the urine.

Activity

The absorption of vitamin B$_{12}$ is dependent upon a gastric mucoprotein from the cardia and fundus of the stomach designated 'the intrinsic factor' by Castle. This appears to interact with vitamin B$_{12}$ in the presence of calcium to protect it during its transit to the ileum. Absorption of vitamin B$_{12}$ takes place in the ileum apparently aided by a specific binding carrier protein. Vitamin B$_{12}$ is stored in the liver and carried in the blood on two proteins (transcobalamin I and II) of which the latter is physiologically more important.

The physiological role of vitamin B$_{12}$ (as cobamide coenzyme) is closely related to that of folacin. It participates in the biosynthesis of labile methyl groups perhaps by cycling 5-methyltetrahydrofolate back into the folate pool. By its interaction with folacin it holds a vital role in the biosynthesis of the purine and pyramidine bases which represent essential constituents of the nucleic acids; in the biosynthesis of methionine and choline and in the conversion of proprionate to succinate. Another important function of vitamin B$_{12}$ in intermediary metabolism is keeping the sulphydryl groups of enzymes in a reduced form. In certain of these reactions it appears that vitamin B$_{12}$ is utilized in coenzyme form with an adenine nucleotide replacing the CN group.

Although the classical signs of vitamin B$_{12}$ deficiency are seen in the haemopoietic and central nervous systems it is clear that it holds an important role in the metabolism of many tissues.

Clinical manifestations of deficiency

As with folacin deficiency, deficiency of vitamin B$_{12}$ produces a megaloblastic maturation of the red cells in the marrow (Plate 13) with a resulting macrocytic anaemia, and a leukopenia with hypersegmented polymorphonuclear leukocytes (Plate 14). A psychosis, sometimes without the blood and bone marrow signs may also rarely occur.

The striking and important difference between vitamin B$_{12}$ and folacin deficiency is that the former, but not the latter, can be associated with subacute combined degeneration of the spinal cord (Figure 34) (degeneration of the dorsal and lateral columns) and this is made worse by the administration of folic acid. It is, therefore, important to establish the diagnosis accurately. On occasions a previously absent subacute combined degeneration of the cord can be precipitated by the administration of folic acid, based upon a wrong diagnosis of the deficiency.

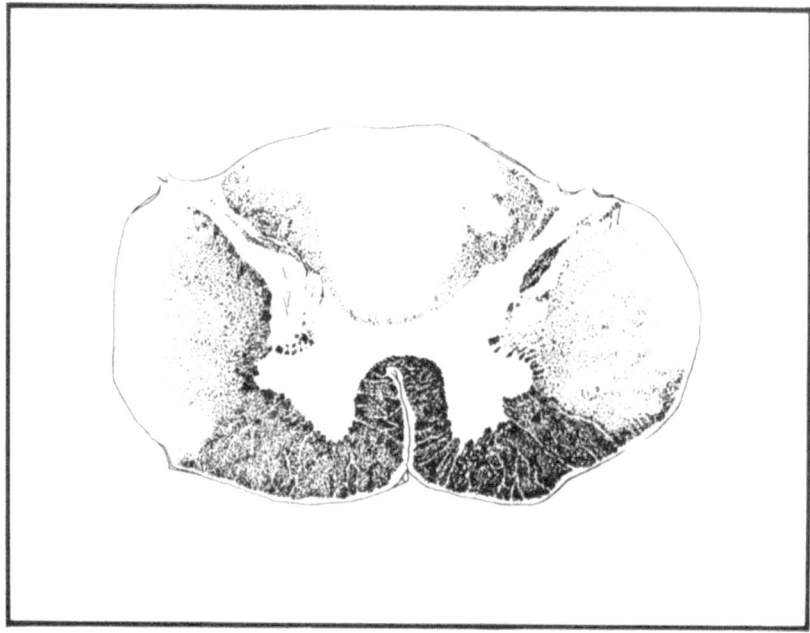

Figure 34   Transverse section of the spinal cord in a case of subacute combined degeneration of the cord

Therapy

Deficiency states

During the relapse stage of pernicious anaemia intramuscular doses of vitamin B_{12} at the rate of $10-20\,\mu g$ per day should be given. It is often convenient to give a large dose infrequently rather than to give a small daily dose. Larger doses of vitamin B_{12} by mouth have been used by some workers, but since absorption in such patients is variable and poor, diffusion probably only permits about 1% of high administered doses to be absorbed, this route cannot be recommended. The first manifestation of a successful response to vitamin B_{12} consists of an intense feeling of well-being and an increase in the appetite, and this often appears within the first 2 days of therapy before any blood changes are seen. In patients who have been treated successfully with vitamin B_{12} for pernicious anaemia the maintenance requirement of approximately $1\,\mu g$ may conveniently be given as an intramuscular dose of about $30\,\mu g$ at monthly intervals.

Other disorders

The general 'tonic' effect of vitamin B_{12} during the early stages of treatment of pernicious anaemia has led to the administration of this vitamin, either orally or parenterally, as a tonic in patients with no evidence of pernicious anaemia. Despite numerous negative controlled trials this form of therapy is still popular and at least has the merit that side effects are virtually unknown. Vitamin B$_{12}$ is also used (despite the lack of scientific proof of its value) in various neuropathies that are unrelated to subacute combined degeneration.

Safety

There are no known adverse effects from single oral doses as high as 100 mg, and doses of 1 mg (300 times the RDA) at weekly intervals for periods up to 5 years are also clear of side effects.

26

Folic acid

Alternative names: folacin, vitamin M or B_c, lactobacillus factor

Folic acid is the parent compound (pteroylglutamic acid – PGA – Figure 35) of a group of several naturally occurring compounds with similar activity designated by the group name folacin. It is a combination of the pteridine nucleus, p-amino benzoic acid and glutamic acid. Replacement of the glutamic acid destroys the vitamin activity. It is stable to heat in neutral and alkaline solution, but unstable in acid solution and to light.

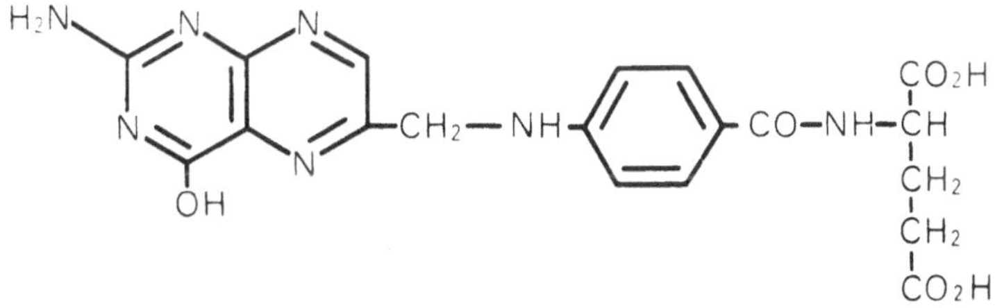

Figure 35   Structural formula of folic acid

Sources

Folic acid is present in liver and kidney and also in brassica, but cooking may reduce the content significantly and the availability from vegetable sources is in any case variable.

Requirements

The US RDA is shown in Table 7 and most countries advise similar levels. The supplement during pregnancy doubles the normal requirement, and folic acid deficiency in pregnant women is common.

Causes of deficiency

Folic acid deficiency is one of the more common of the single vitamin deficiency states and the main causes are shown in Table 53. It is seen fairly commonly in the tropics as a result of reduced dietary intake. It is also encountered in geriatric patients and in chronic alcoholics, in both of which groups there is a combination of a poor diet and reduced absorption from the alimentary canal. Several anticonvulsants can interfere with the folic acid status and lead to a deficiency.

Table 53 The main causes of folic acid deficiency

Cause	Examples
Dietary lack	tropical diets alcoholism the elderly
Increased needs	pregnancy lactation
Poor absorption	malabsorption syndromes alcoholism the elderly
Poor utilization	alcoholism
Drug interference	anticonvulsants barbiturates methotrexate

Another important cause of deficiency is pregnancy. The developing fetus makes considerable demand on the maternal folic acid stores. If these are low at the start of pregnancy the stores are rapidly depleted.

Assessment of vitamin status

In folic acid deficiency the blood and marrow picture are identical to those seen in pernicious anaemia (see Plates 13 and 14). The serum and red cell folic acid may also be used for assessment of the folic acid status as may the histidine test.

Activity

Folic acid in the free form, pteroyl-monoglutamic acid, is mainly absorbed in the proximal part of the intestine by an energy dependent process. There are small liver stores. In the liver folic acid is converted into various metabolically active derivatives of which the main active component is tetrahydrofolic acid (folinic acid) with a formyl group at position 5. The reaction is enhanced by ascorbic acid. The function of folinic acid is to act as a carrier of the 'one carbon' moiety in active form for methylation reactions. The 'one carbon' moiety may be in the metabolically interconvertible forms of formyl (-CHO), active carboxyl (-COOH) or hydroxymethyl (-CH$_2$OH). For the formylation of glutamic acid in the breakdown of histidine the active form is at position 5, for all other reactions it is bound between positions 5 and 10 on tetrahydrofolic acid.

The sources of the carbon moiety, which can be carried direct by folinic acid or indirectly via cobalamin, are shown in Table 54. They are used for several important reactions including: the source of carbons 2 and 8 in purine nucleus synthesis, conversion of glycine to serine, methylation of homocysteine to methionine, synthesis of thymine from uracil and the synthesis of choline.

Table 54 Sources of 'one carbon' moiety carried directly or indirectly by folinic acid

Direct carriage	Indirect carriage
Histidine via formamino glutamate	methionine
The β-carbon of serine	choline via betaine
Glycine	thymine
Acetone and methanol	

There are several folic acid antagonists, including analogues with an amino group substituted for the hydroxyl group in position 4 of the pteridine nucleus. Aminopterin (4-amino folic acid) is the most extensively studied and has been used successfully in the treatment of leukaemia particularly in children (page 174).

Clinical manifestations of deficiency

Folic acid deficiency may occur either as a result of deficient intake, increased requirements (e.g. pregnancy) or deficient absorption (e.g. idiopathic steatorrhoea). The clinical manifestations of folic acid deficiency resemble in many respects those of vitamin B$_{12}$ deficiency, but they should be distinguished because there is one striking difference between the two, namely that folic acid deficiency does not lead to subacute combined degeneration of the cord.

Folic acid deficiency (and cobalamin deficiency) produces a megaloblastic red cell maturation in the bone marrow, a macrocytic anaemia, a leukopenia with hypersegmented polymorphonuclear leukocytes. Deficiency of either of these vitamins (i.e. folic acid or cobalamin) can produce a psychosis with mental deterioration but without blood or bone marrow signs.

Therapy

Deficiency states

When folic deficiency occurs, daily doses of 5–15 mg orally are normally adequate for its relief. The response is usually rapid, and it is rare for parenteral therapy to be required. In some cases of gastrointestinal malabsorption syndromes maintenance therapy with folic acid at a dose of 5–10 mg daily is necessary.

Particular caution must be exercised when there is any suspicion that a macrocytic anaemia or the nervous manifestations are due to pernicious anaemia. In such patients folic acid will precipitate or exacerbate the degeneration of the nervous system, even though it may initially improve the blood picture. In such patients cobalamin is the drug that should be used (see page 168).

Safety

There have been rare reports of gastrointestinal disturbances with high doses of folic acid, and such doses may also mask a diagnosis of pernicious anaemia.

Folate derivative

In the early days of development of substances which are closely related to the vitamins, it was found that antivitamin activity could be developed as opposed to increased vitamin activity, and therapeutic possibilities were hypothesized for such substances. In practice, though, they have been interesting experimentally, therapeutic benefit has been rare, the main exceptions being those related to folate action. Folate plays an important role in the production of purines, and methotrexate and folinate can both inhibit these reactions, though acting at different stages in the metabolic process (Figure 36). These two substances have been found to be valuable in the management of patients with neoplastic disorders, particularly those affecting the blood forming tissues. Their use should be restricted to specialized units, particularly since they have unpleasant side effects.

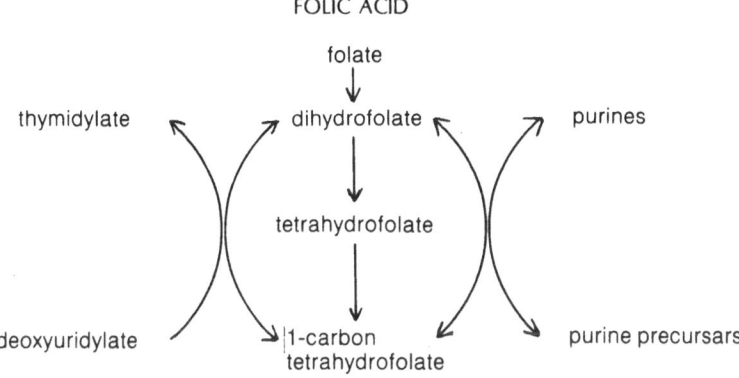

FOLIC ACID

X = inhibition by methotrexate

Figure 36   Mode of action of methotrexate as an antimetabolite

Folinic acid (page 173) is now available for the relief of toxic reactions caused by methotrexate.

27

Pantothenic acid

Alternative name: vitamin B_5

The structural formula of pantothenic acid is shown in Figure 37, the dextrarotatory being the active form. The free acid is a pale yellow viscous oil soluble in water and alcohol and unstable to acid, alkali and heat.

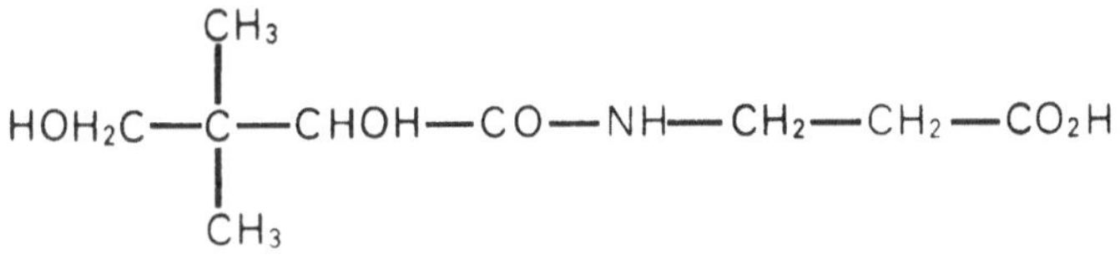

$$HOH_2C-\underset{\underset{CH_3}{|}}{\overset{\overset{CH_3}{|}}{C}}-CHOH-CO-NH-CH_2-CH_2-CO_2H$$

Figure 37 Structural formula of pantothenic acid

Sources

Pantothenic acid, as its name implies, occurs widely in all animal and plant tissues.

Requirements

Few countries offer advice about the pantothenic acid requirement, and the USA authorities only express a range (4–7 mg daily).

Causes of deficiency

A deficiency of pantothenic acid can occur as part of a severe general nutritional deficiency, but other causes are unknown.

Assessment of vitamin status

The pantothenic acid status can be assessed from the blood level and by determining the acetylation ability.

Activity

Pantothenic acid, its salts and its alcohol are absorbed well from the intestinal tract and converted in the tissues into pantotheine, which by combination with pyrophosphate, D-ribose-3-phosphate and adenine forms coenzyme A (CoA). This coenzyme is present in all tissues and is very important for all intermediary metabolism (Figure 38). Coenzyme A acts as a transfer mechanism for carboxylic acids, and such acids when bound to CoA have a high potential for transfer to other groups – 'active' state. The most important of these reactions is to form 'active' acetate capable of further important reactions, including catabolic participation in the Kreb's cycle and anabolic reactions leading to, for example, the steroids (Figure 38). Coenzyme A can also act as the carrier for succinic acid derived from the decarboxylation of ketoglutaric acid in the Kreb's cycle and used in its 'active' form for various anabolic reactions. The combination of the carboxyl groups with coenzyme A occurs at the terminal sulphydryl group of pantotheine.

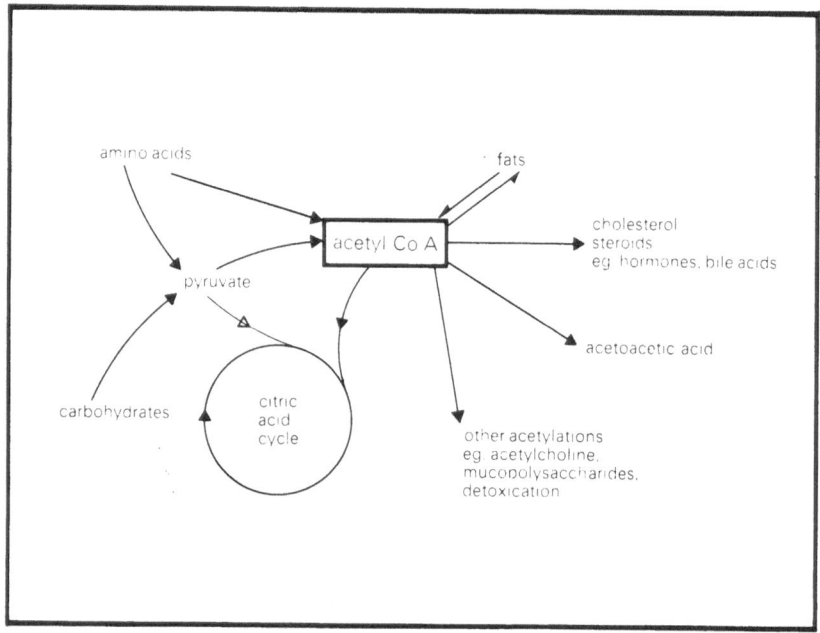

Figure 38 The important role of coenzyme A in metabolism

Some pantothenic acid analogues including ω-methyl pantothenic acid can act as antagonists.

Clinical manifestations of deficiency

Although symptoms related to general malaise have been produced in volunteer depletion studies, there is no evidence that a natural deficiency state in humans gives rise to similar symptoms. The only deficiency disorder is the so-called burning feet syndrome, characterized by paraesthesia and circulatory disturbances in the legs.

Therapy

Deficiency states

The only manifestation which may relate to pantothenic acid deficiency is the 'burning feet' syndrome, and patients in the tropics complaining of this disorder have been treated with pantothenic acid or the related alcohol with variable benefit.

Other disorders

It has been claimed that panthenol (the alcohol related to pantothenic acid) may improve the healing of bed sores and varicose ulcers, but the claim does not stand up to critical analysis.

Many studies have been undertaken with parenteral pantothenic acid (200−600 mg per day) in the prevention and treatment of paralytic ileus. The results remain equivocal, but on balance the use of this therapy is probably justified as one component of the overall treatment.

Safety

This vitamin and the related panthenol are very well tolerated in doses in excess of 1000 times the RDA, and a dose which is unsafe for man has not yet been defined.

28

Biotin

Alternative names: vitamin H, vitamin B$_8$

Biotin is a cyclic derivative of urea with an attached thiophene ring. Of the eight possible isomers, only D-biotin (Figure 39) has vitamin activity. It exists as fine colourless crystals soluble in water and alcohol and stable to heat, acids and alkalis.

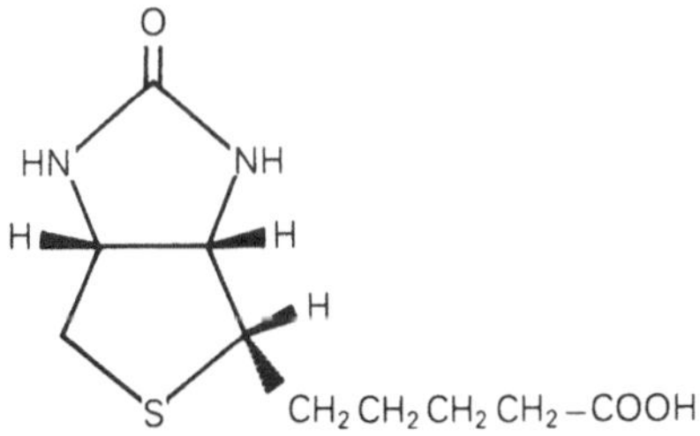

Figure 39   Structural formula of biotin

Sources

Biotin is widely distributed in low concentration in both animal and vegetable foodstuffs, higher values being found in yeast, liver and kidney.

Requirements

The situation for biotin is similar to that for pantothenic acid with little evidence on which advice about daily levels can be based. The USA RDA suggests the range 0.1–0.2 mg daily for adults, although the quality of the evidence is poor.

Causes of deficiency

With the exception of infancy and during haemodialysis and total parenteral nutrition without biotin, a natural deficiency of biotin is excessively rare in humans. Biotin deficiency has recently been reported after long term anticonvulsant therapy but the clinical significance of this is not yet known.

Assessment of vitamin status

The biotin status may be assessed from the plasma level.

Activity

Biotin is readily absorbed from the intestinal tract and widely distributed in all tissues. It takes part in numerous carboxylation reactions, labile carboxybiotin yielding an 'active' form of carbon dioxide. The main reactions in which biotin is involved include carboxylation of pyruvic acid to oxaloacetic acid, conversion of propionic acid to succinic acid via methyl malonyl CoA; transcarboxylation in the catabolism of certain amino acids, conversion of acetyl CoA to malonyl CoA in the formation of long chain fatty acids and a role in pyridine and pyrimidine synthesis.

Clinical manifestations of deficiency

Volunteer depletion studies produce a mild dermatitis and anaemia but no *natural* adult deficiency disorder is known (but note above, the circumstances during which it can occur). Two related conditions in infants, seborrhoeic dermatitis and desquamative erythroderma (Leiner's disease) are probably related to biotin deficiency. There is one rare disorder with biotin dependency in which an abnormality in the carboxylation of propionic acid leads to a ketotic hyperglycinaemia.

Therapy

Deficiency states

The only natural disorders attributable to biotin deficiency in humans are seborrhoeic dermatitis and desquamative erythroderma (Leiner's disease) in infants. In such infants biotin 2–5 mg daily by mouth or injection often leads to a spectacular improvement in the general condition and regression of the skin lesions.

182

Safety

At doses even substantially above those for the maintenance of tissue integrity biotin is without side effects.

29

Vitamin C

Alternative names: ascorbic acid, antiscorbutic vitamin

Vitamin C is the general designation which is applied to the levo-enolic form of 3-keto-l-glucofuranolactone (Figure 40). Vitamin C is water soluble to form a fairly strongly acid solution (pH 3 at 0.5%). In the presence of water, vitamin C is rapidly oxidized. At a second stage of the oxidation, an irreversible metabolite is formed.

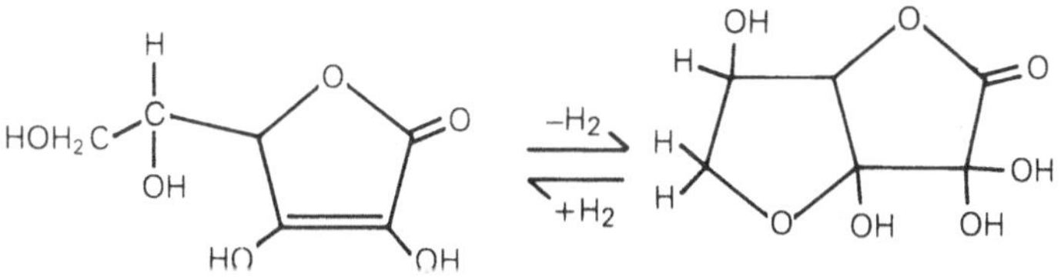

Figure 40 Structural formulae of ascorbic acid and dehydroascorbic acid which form a redox system

Sources

Vitamin C is widely distributed in high concentrations particularly in the fleshy parts of citrus fruits and vegetables. The exact content shows great variation depending on the degree of ripeness, storage conditions, etc.

Requirements

The daily requirements are still a matter of considerable dispute and vary markedly from one country to another and from one author to another. This depends in the main on the criteria that are used for the determination.

The figures published in the NRC, USA are shown in Table 7. Supplements are advised for pregnant and nursing women.

The author shares the view held by many current workers in the field that a value for adults of at least 100 mg is more realistic, particularly if we take account of the effect of smoking (indulged in by over 50% of the population – see page 15 – e.g. France advises 120 mg for smokers); protection against nitrosamine formation (page 14) and potentiation of iron absorption from the diet.

An interesting recent study in the elderly has added further weight to the view that the current recommended intake is too low. At least 60 mg is needed daily to maintain reasonable levels, while the plasma vitamin C level plateaus in this group at about 100 mg per day. (Figure 41.)

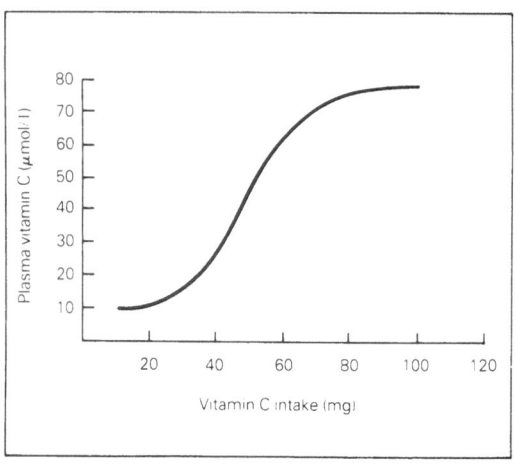

Figure 41 Plasma vitamin C levels in elderly women given various daily intakes of vitamin C. (Based on H.M.V. Newton et al., Relation between intake and plasma concentration of vitamin C in elderly women, British Medical Journal, **287**, 1429 (1983))

Causes of deficiency

Humans, other primates and guinea pigs are the only mammalian species that do not possess the necessary enzyme for the endogenous synthesis of vitamin C and these animals are, therefore, the only ones that require a dietary source of the vitamin.

Vitamin C is one of the vitamins about which there is still considerable dispute concerning the dietary needs, and hence the extent of the deficiency. A further complicating factor in the determination of the needs is a question of the relevance of subclinical deficiency states. The factors which are known to influence the vitamin C needs are detailed on page 191.

VITAMIN C

Table 55 Percentage proportion of adult population groups in the UK and Australia with low leukocyte C reserves

Group	UK Biochemical Deficient*	UK Scorbutic Deficient**	Australia Biochemical Deficient*	Australia Scorbutic Deficient.**
Healthy with good intake	—	—	nil	nil
Young healthy	3	nil	—	—
Apparently healthy affluent group	—	—	11	4
Elderly healthy	20	3	—	—
Socioeconomic stress group	—	—	79	50
Patients with diseases	—	—	45	24
Elderly outpatients	68	20	—	—
Patients with malnutrition	—	—	69	69
Alcoholics with malnutrition	—	—	100	100
Institutionalized elderly	95	50	—	—
Institutionalized young	100	30	—	—

* These showed depleted plasma ascorbate levels
** These showed plasma ascorbate levels at scorbutic levels
Based on Scharah (1981), and Nobile and Woodhill (1981)

187

Surveys indicate that tissue stores of vitamin C are rapidly depleted particularly by infections, and hence any of the normal causes of poor nutritional intake may give rise to a deficiency in such circumstances. The main groups at risk (Table 55) in industrially developed countries are the elderly and those who live alone (including young adults). Since vitamin C is found in abundance in citrus fruits which grow in the tropics, scurvy is one of the vitamin deficiencies which is relatively rare in many Third World countries (except those with severe drought conditions).

Assessment of vitamin status

The vitamin C status can be determined from the plasma or urine level. A more reliable estimate of the tissue status is afforded by the buffy layer determination.

Activity

Unlike most species of animals, the human (together with other primates and the guinea pig) is unable to synthesize vitamin C and must rely on exogenous sources. Vitamin C is normally readily absorbed from the intestinal tract but achlorhydria reduces the extent of the absorption. Unlike most water soluble vitamins, limited stores are held in the body. Whether the high level in the liver and adrenal cortex represent stores or active vitamin is not clear, but it appears that in the adrenal glands it is metabolically active since it is depleted by stress or the administration of adrenocorticotrophic hormone. Vitamin C is excreted in the urine, sweat and faeces, the main loss being in the urine. Active tubular reabsorption has been demonstrated to a tubular maximum rate. In addition to urinary excretion in unchanged form there is catabolism and excretion as oxalic acid and 2,3-diketogulonic acid.

The enediol groups at the second and third carbon atom are sensitive to oxidation, and can easily convert into a diketo group forming dehydro-ascorbic acid. Within the body this reaction may perform an important function as a redox system probably in association with glutathione and cysteine.

Knowledge is still incomplete about the biochemical and physiological role of vitamin C in the body. The reactions that may be relevant are shown in Table 56. For some of these diverse effects it is suggested that there may be a link with cyclic AMP and cyclic GMP.

In more general terms it is known that vitamin C is required for the elaboration of intercellular cement substance and hence for tissue repair.

A recent study has indicated that the administration of vitamin C increases the physical working capacity of young adolescents as determined by their maximum oxygen utilization. This association was more

Table 56   Some of the principal functions of vitamin C

Cofactor for mixed function oxidases
    e.g. propyl hydroxylase
        dopamine hydroxylase
        cholesterol hydroxylase, e.g. in formation of glucocor-
        ticoids
        amino acid hydroxylases

Conversion of folic acid to folinic acid

Cofactor for metal ion metabolism
    e.g. iron transport and mobilization
        copper metabolism

Free radical scavenger
    including: direct reactions with free radicals
               indirect reactions through vitamin E

Stimulates leukocyte phagocytic activity

Assists antibody formation

pronounced in those subjects having low plasma vitamin C levels before treatment. A maximum effect was found when the plasma vitamin C reached the level of 0.8–0.9 mg/dl. On the basis of these data it may be concluded that the optimum aerobic capacity is associated with the minimum daily vitamin C intake of 80–100 mg which would provide this plasma vitamin C level (Figure 42).

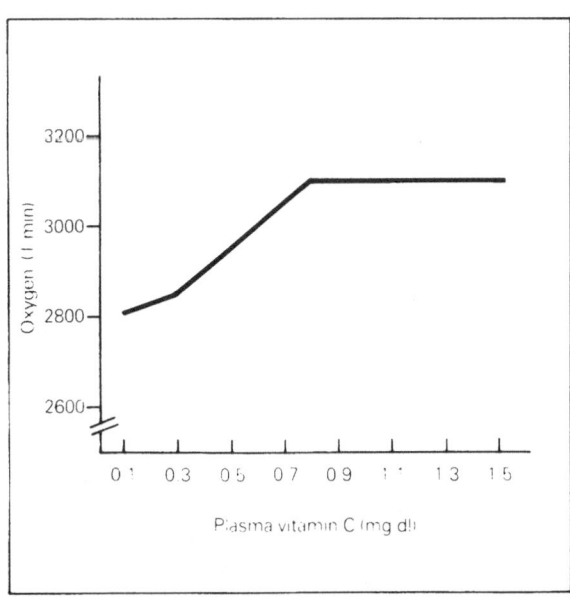

Figure 42   Influence of plasma vitamin C level on oxygen utilization

Interference of ascorbic acid with laboratory tests

Since the vitamins are reactive substances it is surprising that their presence in body fluids produces very little interference with laboratory examinations for other substances. Indeed the only member of the series that produces problems is vitamin C.

When vitamin C has been consumed in relatively large amounts it can interfere with the analysis of glucose, uric acid, creatinine and inorganic phosphate in the urine, but even in megadosage no interference is found within the serum.

Vitamin C consumed in amounts greater than 1 g per day can also interfere with the detection of occult blood.

Clinical manifestations of deficiency

Insufficient intake of vitamin C results in scurvy, a disease characterized by multiple haemorrhages. In adults the earliest symptoms are lassitude, weakness and muscle pains followed by bleeding gums (Plate 15), gingivitis and loosening teeth which are the first recognized clinical signs (but note that these are not present in edentulous people). At this stage hyperkeratotic follicular papules may occur on the calves and buttocks with spiral and unerupted hairs, but they are usually only recognized after the diagnosis has been made on the basis of other signs.

These early signs are followed by minute haemorrhages under the skin particularly in sites of stress and trauma. In more severe cases extensive haemorrhages occur either into the tissues (Plates 16–18), e.g. the muscles, retina etc., or through membranes, e.g. the intestinal tract. One interesting recent discovery is that the vitamin C level correlates inversely with cerebrovascular disease (particularly cerebral haemorrhage in younger subjects). This observation needs further confirmation before vitamin C can be advised prophylactically.

Infantile scurvy usually occurs between the ages of 6 and 18 months. The infant is irritable and cries on being handled. Although haemorrhages may occur anywhere in the body, common sites are under the periosteum of long bones or at the costochondral junction – producing the 'scorbutic rosary'.

Neurolathyrism is a disorder which has been recently recognized as being dependent on vitamin C deficiency. The consumption of Khesava (the legume *Lathyrus sativa*) in India and Bangladesh leads to a neurological disturbance characterized by paralysis, tremor and ataxia. Normal plasma vitamin C levels prevent the disorder and the administration of high doses of vitamin C are curative in established neurolathyrism.

Therapy

Deficiency states

Vitamin C is specific in the treatment of scurvy and subscorbutic conditions. The dosage recommended has varied from 200 – 2000 mg daily. On balance there would appear to be merits in treating such patients with relatively high doses (say 500 – 1000 mg daily) until the tissue reserves are saturated.

Other disorders

Since wound healing is delayed and haemorrhagic manifestations are common in scurvy, there is medical opinion favouring the administration of vitamin C in a variety of miscellaneous medical states. It is also known that stress leads to vitamin C depletion which in turn reduces adrenal cortical function so that there is some additional rationale for the use in some of these disorders. The conditions involved are:

(1) *Anaemia*. The association of anaemia with scurvy has long been recognized. Vitamin C improves iron absorption and, therefore, it may be given with benefit in patients with iron deficiency anaemia.

(2) *Haemorrhagic disorders*. In those patients with haemorrhagic disorders who show laboratory evidence of increased capillary fragility, vitamin C may be administered as a proportion of patients respond dramatically.

(3) *Infectious diseases*. Controlled studies have shown that in the course of various infections the vitamin C in the body is depleted and urinary excretion is considerably reduced. This was particularly noted in active tuberculosis where the tissue level was shown to be proportional to the severity of the disease. Maintenance of the vitamin C levels at tissue saturation is recommended on general grounds. There is still no firm consensus about the merits of large doses (1 – 3 g daily) in aborting the common cold. In common with many doctors the author feels justified in using it for himself and his family, particularly if the vitamin C can be taken as soon as the first manifestation of the cold (the slight dry sore throat) is experienced. This may be due to an influence on the immune mechanism.

(4) *Gastrointestinal disturbances*. In such conditions vitamin C deficiency may arise through impaired absorption, and higher intake is desirable. This is particularly true in patients who have been treated with 'ulcer diets' which usually render them vitamin C deficient. Most surgeons routinely administer vitamin C in such patients

before surgery to reduce the risk of excessive bleeding and to improve wound healing.

(5) *Surgery.* Since vitamin C deficiency adversely affects the formation of collagen, thus delaying or preventing healing, increased intake of vitamin C is desirable in any form of surgery. This is particularly true if there has been considerable trauma which may have further depleted the vitamin C levels.

(6) *Fractures.* Evidence exists that fractures may themselves be a sign for vitamin C administration in that callus formation is improved when vitamin C levels are raised. It is, therefore, common practice for vitamin C to be given at the same time as calciferol for fractures, particularly in the elderly.

(7) *Dental and oral conditions.* Vitamin C given orally before and after dental extraction results in more rapid healing of the gum tissue with absorption of the alveolar bone margins. Vitamin C has also been administered to those with spongy bleeding gums and loosened teeth, even without clear evidence of scurvy.

(8) *Infants.* Many artificial milk diets are not dependable sources of vitamin C and in consequence, when an infant dislikes or poorly tolerates fresh fruit juices, vitamin C may itself be substituted.

(9) *General hospital care.* Many patients maintained for long periods on hospital diets (e.g. geriatric or psychiatric wards) show evidence of depleted vitamin C reserves. In such patients there are merits in supplementing the vitamin C levels on grounds of general health.

(10) *Gastritis and pernicious anaemia.* In such patients there is clear evidence of increased formation of nitrosamines in the intestinal tract from ingested nitrates. Since such nitrosamines may be carcinogenic, high intake of vitamin C divided regularly over the 24 hour period has been shown to reduce the risk of nitrosamine formation.

(11) *Smoking.* There is now clear evidence that cigarette smoking depletes the vitamin C body levels (page 15). Therefore, patients who continue to be heavy smokers should be supplemented with vitamin C on general health grounds (page 15). A daily intake in excess of 100 mg is advised.

(12) *Cystine stone formation.* Vitamin C can be used prophylactically in this rare disorder.

(13) *Neurolathyrism.* Vitamin C is protective against neurolathyrism (page 190) and higher doses (55–1000 mg per day) can reverse the established disease.

Safety

Vitamin C is one of the vitamins which is often consumed in larger quantities. Yet an extensive and very thorough analysis of the data published recently has shown a substantial level of safety though some patients, particularly in the early days of high dose administration do experience a laxative effect thought to be due to a local irritant action. However, even this mild and harmless adverse effect is not found consistently.

Oxalate is the main metabolite of vitamin C, and it has been postulated that this could lead to the formation of kidney oxalate stones. The conversion of vitamin C to oxalate is limited, and does not in fact reach critical levels even after doses of vitamin C as high as 10 g daily. Even in those with inborn metabolic errors, no stones are seen despite high levels of oxalate excretion.

Other reputed adverse effects that have not been substantiated in recent studies are disturbed electrolyte balance, increased red cell lysis, decreased immunological tolerance, mutagenicity, rebound scurvy and suppression of cobalamin activity. It is important to understand that vitamin C itself is a reactive substance in the redox system, and in consequence can give rise to false reactions, particularly in certain analytical tests involving a colour response. This has been reported for the analysis of glucose, uric acid, creatinine and occult blood.

Vitamin C enhances iron absorption. Though there are no reports of excessive iron absorption attributable to large intakes of vitamin C in *normal* people, in the rare cases of excessive gastrointestinal iron absorption the use of high doses of vitamin C is contraindicated.

Overall the exhaustive recent review of the scientific data concluded that vitamin C is a safe substance even with an intake up to 10 g per day.

REFERENCES

References for surveys

AJAYI, O.A. (1981). Riboflavin status of Nigerian teenagers: an enzymatic reassessment. In: Howard, A.N. and Baird, I.McL. (eds). *Recent Advances in Clinical Nutrition.* pp. 24–6, J. Libbey, London

ARROYAVE, G., AGUILAR, J.R., FLORES, M. and GUZMAN, M.A. (1979). Evaluation of sugar fortification with vitamin A at the national level. *Inst. of Nutrition of Central America and Panama, Scientific Publication No. 384*

BARNES, M. (1983). The usefulness of the measurement of red blood cell enzyme activation in the detection of vitamin B_1, B_2 and B_6 deficiency. In: Hanck, A. (ed). *Vitamins in Medicine.* p. 69 *et seq.* H. Huber, Berne

BAYONINI, R.A. and BOSALKI, S.B. (1976). Evaluation of methods of coenzyme activation of erythrocyte enzymes for detection of deficiency of vitamins B_1, B_2 & B_6. *Clin. Chem.,* **22,** pp. 327–35

BRIN, M., DIBBLE, M.V., PEEL, A., MCMULLEN, E., BOURQUIN, A. and CHEN, N. (1965). Some preliminary findings on the nutritional status of the aged in Onondaga County, New York. *Am. J. Clin. Nutr.,* **17,** pp. 240–58

BRUBACHER, G., BRUN, P. and MAINGUY, P. (1978). Compte-rendue d'une enquete alimentaire effectuee dans la region parisienne sur des jeunes adultes et des femmes enceintes de 1968–1970. Documentation Roche, Paris

BRUBACHER, G. (1979). Methodisches zu Ernahrungserhebungen. In: *Probleme um Ernahrungserhebungen.* Symposium der Oesterreichischen Gesellschaft fur Ernahrungsforschung, Wien, pp. 67–76

CAMPBELL, G.A., HOSKING, D.J., KEMM, J.R. and BOYD, R.V. (1984). How common is osteomalacia in the elderly? *Lancet,* **2,** pp. 386–8

CHANARIN, I. and ROTHMAN, D. (1971). Further observations on the relation between iron and folate status in pregnancy. *Br. Med. J.,* **2,** pp. 81–4

CHAVEZ, A., MATA, A. and SANDOVAL, J. (in press). Possibilities of enriching sugar with micronutrients in Mexico. *Int. J. Vitam. Nutr. Res.,* Suppl. 27

DHEW (1979). Caloric and selected nutrient values for persons 1–74 years of age. *USA Vital. & Health Statistics: Series 11,* No. 209 DHEW Publ. No. (PHS) 79–1659

DOSTALOVA, Quoted by Hauser, G.A. (in press). Vitamin requirements in pregnancy. *Int. J. Vitam. Nutr. Res.,* Suppl. 27

EDDY, T.P. (1971). A study of the relationship between the Hess test and leucocyte ascorbic acid levels in a clinical trial. *Br. J. Nutr.,* **27,** pp. 537–42

ELWOOD, P.C., SHINTON, N.K. and WILSON, C.I.D. (1971). Haemoglobin, vitamin B_{12} and folate levels in the elderly *Br. J. Haematol.,* **21,** pp. 557–63

EXTON-SMITH, A.N. and SCOTT, D.L. (1968). *Vitamins in the Elderly.* Wright, Bristol

GARN, S.M. and CLARK, D.C. (1975). Nutrition, growth, development and maturation finding from Ten State Nutrition survey. *Pediatrics,* **56,** pp. 306–19

GOEL, L. In: Arneil, G. (ed). (1979). *The Importance of Vitamins to Human Health.* MTP Press, London

197

HARJU, A. (1979). *Proceedings of 3rd European Nutrition Conference*, Uppsala. p. 118

HELLER, S. (1974). Vitamin B_1 status in pregnancy. *Am. J. Clin. Nutr.*, **27**, pp. 1221–2

HOORN, R.K.J., FLIKWEERT, J.P. and WESTERINK, D. (1975). Vitamin B_1, B_2 & B_6 deficiencies in geriatric patients measured by coenzyme stimulation of enzyme activities. *Clin. Chim. Acta*, **61**, pp. 151–62

HOTZEL, M.S. (1979). *Proceedings of 3rd European Nutrition Conference*, Uppsala. p. 126

HURDLE, A.D.F. and WILLIAMS, T.C.P. (1966). Folic acid deficiency in elderly patients admitted to hospital. *Br. Med. J.*, **2**, pp. 202–5

INCAP/CDC-Nut. Prog. Nutritional evaluation of the population of Central America and Panama: Regional Summary., INCAP/CDD-Nutrition Program. DHEW Publication No. HSM-72-8120 (1972). *US Department of Health, Education and Welfare*, Washington DC

LABADARIOS, D. and ROSSOUW, J.E. (1981). Wanvoeding in drie hospitaalbevolkings. *S. Afr. Med. J.*, **60**, pp. 213–16

LEMOINE, A., LE DEVEHAT, C., CODACCIONI, J.L., MONGES, A., BERMOND, P. and SALKELD, R.M. (1980). Vitamin B_1, B_2, B_6 & C status in hospital inpatients. *Am. J. Clin. Nutr.*, **33**, pp. 2595–600

LOEWENSTEIN, F.W. (1981). Major nutritional findings from the first health and nutrition examination survey in the United States of America, 1971–1974. *Bibl. Nutr. Dieta* No. 30, pp. 1–16. Karger, Basel

McLEAN, H.E., WESTON, R., BEAVEN, D.W. and RILEY, C.G. (1976). Nutrition of elderly men living alone. *N.Z. Med. J.*, **84**, pp. 305–09

MILNE, J.S., LONERGAN, M.E., WILLIAMSON, J., MOORE, F.M.L., MCMASTER, R. and PERCY, N. (1971). Leucocyte ascorbic acid levels and vitamin C intake in older people. *Br. Med. J.*, **4**, pp. 383–7

MONGEAU, E. (1980). Carences et sub-carences en vitamines et mineraux au Canada et dans d'autres pays industrialises. *Proceedings Colloque 'Enrichissement d'Aliments en Vitamines'*, Paris, pp. 4–41

MORA, J.O. (in press). Nutritional status of the Colombian population. *Int. J. Vitam. Nutr. Res.*, Suppl. 27

NNANYOGO, D.O. (1981). A review of nutritional status of children in a Southern Nigerian population, In: Howard, A.N. and Baird, I.McL. (eds). *Recent Advances in Clinical Nutrition*. J. Libbey, London

OWEN, G.M., KRAIN, K.M., GARRY, P.J., LOWE, J.E. and LUBIN, A.H. (1974). A study of nutritional status of preschool children in the United States. *Pediatrics*, **53**, pp. 597–601

POH TAN, S., WENLOCK, R.W. and BUSS, D.H. (1984). Folic acid content of the diet in various types of British households. *Human Nutrition: Applied Nutrition*, **38A**, pp. 17–22

REDDY, V. (1980). *Meeting on Vitamin A Deficiency and Xerophtalmia, Jakarta*

RONCADO, M.J. (1980). In: *Congresso International de Nutricao*, Rio de Janiero, Plenum Press, New York

SCHLETTWEIN-GSELL, D. (1975). Erhebungen uber die Ernahrung von alten Menschen. In: Brubacher, G. and Ritzel, G. (eds). *Zur Ernahrungssituation der Schweizerischen Bevolkerung*. pp. 83–98, H. Huber, Bern.

SMITHELLS, R.W. (1980). Maternal nutrition during pregnancy and lactation. In: Aebi, H.L. (ed). *Maternal Nutrition During Pregnancy and Lactation*. 123, H. Huber, Bern

SOLON, F.S., POPKIN, B.M., FERNANDEZ, T.L. and LATHAM, M.C. (1978). Vitamin A deficiency in the Philippines: a study of xerophthalmia in Cebu. *Am. J. Clin. Nutr.*, **31**, pp. 360–8

STRANSKY, M. (1980). *Mitt. Gebiete Lebensmittel Hygiene*, **71**, pp. 163–81

STRATIGOS, J.D. and KATSAMBAS, A.D. (1982). Pellagra: a reappraisal. *Acta Vitaminol. Enzymol.*, **4**, pp. 115–21

TARWOTJO, H.J. (1980). *Nutritional Blindness Prevention Project, Indonesia. Final Report* July 1980, 3

VARADI, S. and ELWIS, A. (1966). Folic acid deficiency in the elderly. *Br. Med. J.*, **2**, pp. 410–15

VIMOKESANTET, V.M. (1979). *Proceedings of the 3rd European Nutrition Conference*, Uppsala. pp. 126–7

WENGER, R., ZIEGLER, B., KRUSPL, W., SYRE, B., BRUBACHER, G. and PILLAT, R. (1979). Beziehungen zwischen dem Vitaminstatus (Vitamin A, B_1, B_2, B_6, und C), klinischen Bedfunden und den Ernahrungsgewohnheiten in einer Gruppe von alten Leuten in Wien. In: *Probleme um Ernahrungserhebungen, Symposium der Oesterreichischen Gesellschaft fur Ernahrungsforschung*, Wien, pp. 247–62

WILSON, D. and DA SILVA NERY, M.E. (1983). Hypovitaminosis A in Rio Grande de Sol, Brazil. In: Hanck, A. (ed). *Vitamins in Medicine*, pp. 35–44. H. Huber, Berne

Therapeutic index

ACNE – USE OF VITAMIN A
Straumfjord, J.V. (1943). Vitamin A: its effect on acne. A study of one hundred patients. *Northwest Med.*, **42**, 219

ACNE – USE OF RETINOIC ACID
Heel, R.C. (1977). Vitamin A acid: a review of its pharmacological properties and therapeutic use in the topical treatment of acne vulgaris. *Drugs*, **14**, (6), 401

ALCOHOLISM
Thomson, A.D. (1978). Alcohol and nutrition. *Clin. Endocrin. Metab.*, **7**, 405

ANOSMIA
Duncan, R. and Briggs, M. (1962). Treatment of uncomplicated anosmia by vitamin A. *Arch. Otolaryngol.*, **75**, 116

ATROPHIC RHINITIS
Mozumder, A. (1960). Role of vitamin A in the treatment of atrophic rhinitis. *Ind. J. Otolaryngol.*, **XII**, (2), 1

BURNING FEET SYNDROME
Gopalan, C. (1946). The 'Burning-Feet' syndrome. *Ind. Med. Gazette*, p. 22

CARPAL TUNNEL SYNDROME
Ellis, J.M. *et al.* (1982). Response of vitamin B6 deficiency and the carpal tunnel syndrome to pyridoxine. *Prox. Nat. Acad. Sci. USA*, **79**, (23), 7494

CEREBRAL HAEMORRHAGE IN PREMATURE INFANTS
Chiswick, M.L. *et al.* (1983). Protective effect of vitamin E (DL-alpha-tocopherol) against intraventricular haemorrhage in premature infants. *Br. Med. J.*, **287**, 81

CEREBRAL HAEMORRHAGE
Acheson, R.M. and Williams D.R.R. (1983). Does consumption of fruit and vegetables protect against stroke? *Lancet*, **1**, 1191
Vollset, S.E. and Bjelke, E. (1983). Does consumption of fruit and vegetables protect against stroke? *Lancet*, **2**, 742

CHINESE RESTAURANT SYNDROME
Folkers, K. *et al.* (1984). The biochemistry of vitamin B_6 is basic to the clause of the Chinese restaurant syndrome. *Hoppe-Seylers Z., Physiol. Chem.*, **365**, (3), 405

COMMON COLD
Anderson, T.W. (1979). Vitamin C and the common cold. *N.Y. State J. Med.*, **79**, (8), 1292
Basu, T.K. and Schorah, C.J. (1982). In: *Vitamin C in Health and Disease* p. 93. (London: Croom Helm)

CRAMPS
Klein, H.O. (1954). Laktoflavintherapie der Wadenkrampfe in der Schwangerschaft. (The treatment of cramp of the legs during pregnancy with riboflavin). *Zentralbl. Gynaerkol.*, **76**, 344

CYSTINURIA
Lux, B. and May, P. (1983). Long-term observation of young cystinuric patients under ascorbic acid therapy. *Urol. Int.*, 1983, **38**, 91

DUPUTYRENS CONTRACTURE
Thomson, G. (1949). Treatment of Dupuytren's contracture with vitamin E. *Br. Med. J.*, **2**, 1382

ENCEPHALOMALACIA
Harwitt, M.K. and Bailey, P. (1959). Encephalomalacia. *Arch. Neurol. Psychiatry*, **1**, 312

CYSTATHIONURIA
Bartlett, K. (1983). Vitamin-responsive inborn errors of metabolism. *Adv. Clin. Chem.*, **23**, 141

HARTNUP DISEASE
Treatment of Hartnup disease with nicotinic acid. (1984). *Nutr. Rev.*, **42**, (7), 251

HAEMOLYTIC ANAEMIA
Gross, S. (1976). Hemolytic anemia in premature infants: relationship to vitamin E, selenium, glutathione peroxidase, and erythrocyte lipids. *Semin. Hematol.*, **13**, (3), 187

HYPERCHOLESTEROLAEMIA
Carlson, L.A. and Orö, L. (1973). Effect of treatment with nicotinic acid for one month on serum lipids in patients with different types of hyperlipidemia. *Atherosclerosis*, **18**, 1

HYPEREMESIS CIRAVIDARUM
Baum, G. et al. (1963). Meclozine and pyridoxine in pregnancy sickness. *Practitioner*, **190**, 251

INFANCY CONVULSIONS
Heeley, A. et al. (1978). Pyridoxol metabolism in vitamin B_6-responsive convulsions of early infancy. *Arch. Dis. Childh.*, **53**, (10), 794

INFANTILE SEBORRHOEIC DERMATITIS
Nisenson, A. (1957). Seborrheic dermatitis of infants and Leiner's disease: a biotin deficiency. *J. Pediatr.*, **51**, 537

INTERMITTENT CLAUDICATION
Haeger, K. (1974). Long-time treatment of intermittent claudication with Vitamin E. *Am. J. Clin. Nutr.*, **27**, 1179
Marks, J. (1962). Critical appraisal of the therapeutic value of α-tocopherol. *Vitam. Horm. (NY).*, **20**, 573

IRRADIATION SICKNESS
Stoll, B.A. (1962). Radiation sickness: an analysis of over 1,000 controlled drug trials. *Br. Med. J.*, **3**, 507

ISONIAZID NEURITIS
Stott, H. et al. (1963). The prevention and treatment of isoniazid toxicity in the therapy of pulmonary tuberculosis. 2. An assessment of the prophylactic effect of pyridoxine in low dosage. *Bull. WHO*, **29**, 457

LEINER'S DISEASE
Nisenson, A. (1959). Seborrheic dermatitis of infants and Leiner's disease: a biotin deficiency. *J. Pediatr.*, **51**, 537

MAPLE SYRUP DISEASE
Bartlett, K. (1983). Vitamin-responsive inborn errors of metabolism. *Adv. Clin. Chem.*, **23**, 141

MIGRAINE
Atkinson, M. (1944). Migraine headache: some clinical observations on the vascular mechanism and its control. *Ann. Intern. Med.*, **21**, (6), 990

THERAPEUTIC INDEX

NEONATAL RETINITIS

Kretzer, F.L. *et al.* (1984). Vitamin E protects against retinopathy of prematurity through action on spindle cells. *Nature*, **309**, 793

Hittner, H.M. *et al.* (1981). Retrolental fibroplasia: efficacy of vitamin E in a double-blind clinical study of preterm infants. *N. Engl. J. Med.*, **305**, (23), 1365

NEURAL TUBE DEFECTS

Smithells, R.W. *et al.* (1981). Apparent prevention of neural tube defects by periconceptional vitamin supplementation. *Arch. Dis. Childh.*, **56**, (12), 911

Rational use of vitamins. (1984). *Drug Ther. Bull.*, **22**, (9), 33

NEUROPATHY

Schaumburg, H. *et al.* (1983). Sensory neuropathy from pyridoxine abuse: a new megavitamin syndrome. *N. Engl. J. Med.*, **309**, (8) 445

PARALYTIC ILEUS

Frazer, J.W. *et al.* (1959). D-Pantothenyl alcohol in management of paralytic ileus. *J. Am. Med. Assoc.*, **169**, 1047

PEYRONIE'S DISEASE

Pryor, J.P. and Farrell, C.R. (1983). Controlled clinical trial of vitamin E in Peyronie's disease. *Prog. Reprod. Biol. Med.*, **9**, 41

POLYUNSATURATED FATTY ACIDS

Debry, G. (1980). Polyunsaturated fatty acids and vitamin E: their importance in human nutrition. *Ann. Nutr.*, **34**, (2), 337

POST-GASTRECTOMY

Duncan, W.H. (1952). Nutritional disturbances incident to gastrectomy. *J. Missouri St. Med. Assoc.*, **49**, 73

PREMENSTRUAL TENSION

Kerr, G.D. (1977). The management of the premenstrual syndrome. *Curr. Med. Res. Opin.*, **4**, (Suppl. 4), 29

RAW FISH NEUROPATHY

Vimokesant, S. *et al.* (1982). Beriberi caused by antithiamin factors in food and its prevention. In: Sable, H.Z. and Gubler, C.J. (eds.) *Thiamin: Twenty Years of Progress. Ann. N.Y. Acad. Sci.*, **378**, 123

SCHIZOPHRENIA

Denson, R. (1962). Nicotinamide in the treatment of schizophrenia. *Dis. Nerv. Syst.*, **23**, 167

SMOKING

Pelletier, O. (1975). Vitamin C and cigarette smokers. *Ann. N.Y. Acad. Sci.*, **258**, 156

SURGERY

Schwartz, P.L. (1970). Ascorbic acid in wound healing – a review. *J. Am. Diet. Assoc.*, **56**, 497

TRIGEMINAL NEURALGIA

Da Silva, E.E. (1956). A vitamina B_1, nas Neuralgias do Trigemeo ("Vitamin B_1 in Trigeminal Neuralgia"). *Rev. Brasil Odontol.*, **14**, 204

XANTHURENIC ACID DISEASE

Bartlett, K. (1983). Vitamin-responsive inborn errors of metabolism. *Adv. Clin. Chem.*, **23**, 141

Selected further reading

General books and monographs

Alfin-Slater, R.B. and Kritchevsky, D. (eds.) (1980). *Nutrition and the Adult.* Vol. 36. *Micronutrients,* p. 36. (New York: Academic Press)

Babior, B.M. (ed.) (1975). *Cobalamin: Biochemistry and Pathophysiology.* (New York: Wiley Interscience)

Barker, B.M. and Bender, D.A. (1980). *Vitamins in Medicine.* 2 Vols, 4th Edn. (London: Heinemann Medical)

Counsell, J.N. and Hornig, D.H. (1981). *Vitamin C – Ascorbic Acid.* (London: Applied Science Publishers Ltd)

Davidson, S., Passmore, R., Brock, J.F. and Truswell, A.S. (1975). *Human Nutrition and Dietetics,* 6th Edn. (Edinburgh: Churchill Livingstone)

DeLuca, H.E. (1978). *The Fat-Soluble Vitamins.* (New York: Plenum Press)

Hanck, A. (ed.). (1983). *Vitamins in Medicine: Recent Therapeutic Aspects.* (Berne: Huber)

Hanck, A. and Ritzel, G. (eds.) (1977). *Re-evaluation of Vitamin C.* (Berne: Huber)

Howard, A. and McLean Baird, I. (eds.) (1981). *Recent Advances in Clinical Nutrition.* Vol. 1. (London: Libbey)

Howard, A.N. (ed.) (1981). *Nutritional Problems in Modern Society.* (London: Libbey)

Machlin, L.J. (ed.) (1980). *Vitamin E.* (New York: Dekker)

Marks, J. (1975). *A Guide to the Vitamins. Their Role in Health and Disease,* 1st Edn. (Lancaster: MTP Press)

Neuberger, A. and Jukes, T.H. (eds.) (1982). *Human Nutrition. Current Issues and Controversies.* (Lancaster: MTP Press)

Nobile, S. and Woodhill, J.M. (1981). *Vitamin C. The Mysterious Redox-System. A Trigger of Life?* (Lancaster: MTP Press)

Norman, A.W. (1979). *Vitamin D. The Calcium Homeostatic Steroid Hormone.* (New York: Academic Press)

Norman, A.W., Schaefer, K., Herrath, D.V. and Grigoleit, H.G. (eds.) (1979). *Vitamin D: Basic Research and its Clinical Application.* (Berlin: Walter de Greyter)

Rechcigl, M. (ed.) (1978). *CPC Handbook Series in Nutrition and Food.* Vol. 1. (West Palm Beach: CPC Press)

Turner, M.R. (ed.) (1982). *Nutrition and Health.* (Lancaster: MTP Press)

Papers on general vitamin topics

Bender, A.E. (1982). The effects of food processing on the vitamins. In Barker, B.M. and Bender, D.A. (eds.) *Vitamins and Medicine.* Vol. 2, pp. 291–318. (London: Heinemann Medical)

Holtzel, D. and Barnes, R.H. (1966). Contributions of the intestinal microflora to the nutrition of the host. *Vitam. Horm.*, (NY), **24**, 115−71

Shive, W. and Lansford, E.M. (1980). Roles of vitamins as coenzymes. In Alfin-Slater, R.B. and Kritchevsky, D. (eds.) *Nutrition and the Adult*. Vol. 36, *Micronutrients*, p. 1. (New York: Academic Press)

Smithells, R.W., Sheppard, S., Schopah, C.J. *et al.* (1980). Possible prevention of neural-tube defects by periconceptional vitamin supplementation. *Lancet*, **1**, 339−40

Woodhill, J.M. (1979). Assessment of diet. In Hetzel, B.S. and Nobile, S. (eds.) *Human Nutrition in Australia. Int. J. Vitam. Nutr. Res.* Suppl. No. 17

Requirements and recommendations

National Research Council/Food and Nutrition Board. (1980). *Recommended Dietary Allowances*. 9th revised Edn. (Washington, DC: National Academy of Sciences)

Bender, D.A. (1980). Requirements, recommendations and intake. In Barker, B.M. and Bender, D.A. (eds.) *Vitamins in Medicine*. 2 Vols, 4th Edn., pp. 1−41. (London: Heinemann Medical)

DHSS. (1979). *Recommended Daily Amounts of Food Energy and Nutrients for Groups of People in the United Kingdom*. Department of Health and Social Security Report on Health and Social Subjects No 15. Report by the Committee on Medical Aspects of Food Policy. (London: HMSO).

Editorial. (1984). New thoughts on the British diet. *Lancet*, **2**, 143−4

FAO/WHO. (1967). *Requirements of vitamin A, thiamine, riboflavine and niacin; report of a joint FAO/WHO Expert Group, Food and Agriculture Organization and World Health Organization*. (FAO Nutrition Meetings Report Series, No. 41 and World Health Organizational Technical Report Series No. 362) Geneva

FAO/WHO. (1970). *Requirements of ascorbic acid, vitamin D, vitamin B_{12}, folate, and iron*. Food and Agriculture Organization and World Health Organization (World Health Organization Technical Report Series, No. 452 and FAO Nutrition Meetings Report Series No. 47), Geneva

Hegsted, D.M. (1978). On dietary standards. *Nutr. Rev.*, **36**, 33−6

Nutritional status

Baines, M. (1978). Detection and incidence of B and C vitamin deficiency in alcohol-related illness. *Ann. Clin. Biochem.*, **15**, 307

Bamji, J.S. (1976). Enzymic evaluation of thiamin, riboflavin and pyridoxine status of parturient women and their new-born offspring. *Br. J. Nutr.*, **35**, 259

Bender, A.E. (1984). Letters: institutional malnutrition. *Br. Med. J.*, **288**, 92−3

Burr, M.L., Milbank, J.E. and Gibbs, D. (1982). Nutritional status of the elderly. *Age Ageing*, **3**, 71−76

DHEW. (1979). *Caloric and selected nutrient values for persons 1−74 years of age United States, 1971−1974, Vital and Health Statistics*. Series II, Data from the National Health Survey: No 209. DHEW Publ. No. (PHS) 79−1659

DHSS. (1979). *Nutrition and Health in Old Age*. Report on Health and Social Subjects No. 16. (London: HMSO)

Horst, R.L., Shepard, R.M., Jorgensen, N.A. and DeLuca, H.F. (1979). The determination of 24,25-dihydroxyvitamin D and 25,26-dihydroxy-vitamin D in plasma from normal and nephrectomized man. *J. Lab. Clin. Med.*, **93**, 277

Ministry of Agriculture, Fisheries and Food. (1977). Household food consumption and expenditure 1975: with a review of the six years 1970−1975, *Annual Report of the National Food Survey Committee*. (London: HMSO)

Owen, G.M. and Owen, A.L. (1982). Nutritional status of North Americans. In Neuberger, A. and Jukes, T.H. (eds.) *Human Nutrition. Current Issues and Controversies*, pp. 73−86. (Lancaster: MTP Press)

Sauberlich, H.E., Dowdy, R.P. and Skala, J.H. (1974). *Laboratory Tests for the Assessment of Nutritional Status*. (Cleveland, Ohio: CRC Press)

Todhunter, E.N. (1980). Nutrition of the Elderly. In Alfin-Slater, R.B. and Kritchevsky, D. (eds.) *Nutrition and the Adult*. Vol. 36. *Micronutrients*, p. 397. (New York: Academic Press)

General therapeutics papers

Aftergood, L. and Alfin-Slater, R.B. (1980). Oral contraceptives and nutrient requirements. In Alfin-Slater, R.B. and Kritchevsky, D. (eds.) *Nutrition and the Adult*. Vol. 36. *Micronutrients*, p. 367. (New York: Academic Press)

Bauernfeind, J.C., Newmark, H. and Brin, M. (1974). Vitamin A an E nutrition via intramuscular or oral route. *Am. J. Clin. Nutr.*, **27**, 234

Bober, M.J. (1984). Letters: Senile dementia and nutrition. *Br. Med. J.*, **288**, 1234

Copeland, E.M. (1979). Nutritional concepts in the treatment of cancer. *J. Florida Med. Assoc.*, **66**, 373−89

Dickerson, J.W.T. (1981). Vitamins and trace elements in the seriously ill patient. *Acta Chir. Scand.*, **507**, 144−50

Dickerson, J.W.T. (1984). Nutrition in the cancer patient. *J.R. Soc. Med.*, **77**, 309−15

Nichoalds, G.E. *et al.* (1977). Vitamin requirements in patients receiving total parenteral nutrition. *Arch. Surg.*, **112**, 1061−4

Shenkin, A. (1981). Additives in parenteral nutrition. *Acta Chir. Scand.*, **507**, 350−5

Smithells, R.W., Sheppard, S., Schorah, C.J. *et al.* (1980). Possible prevention of neural tube defects by preconceptual vitamin supplementation. *Lancet*, **1**, 339

Interference with laboratory tests

Siest, G. *et al.* (1978). Drug interference in clinical chemistry: studies on ascorbic acid. *J. Clin. Chem. Clin. Biochem.*, **16**, 103−10

Drug − vitamin interactions

Roe, D.A. (1984). Nutrient drug interactions. *Nutr. Rev.*, **42**, 141−54

General papers on safety

Arnich, L. *et al.* (1980). Interactions of fat-soluble vitamins in hypervitaminoses. *Ann. NY Acad. Sci.*, **335**, 109−18

Barness, L.A. (1977). Some toxic effects of vitamin C. In Hanck, A. and Ritzel, G. (eds.) *Re-evaluation of vitamin C*. pp. 23−29. (Berne: Huber)

Bauernfeind, J.C. (1980). *The safe use of vitamin A. A report of the international Vitamin A Consultative Groups (IVACC)*. (Washington: Nutrition Foundation)

Briggs, M.H. (1978). Effect of specific nutrient toxicities in animals and man: tocopherols. In Rechcigl, M. Jr. (ed.) *CRC Handbook Series in Nutrition and Food. Section E; Nutritional Disorders*. Vol 1. *Effect of Nutrient Excesses and Toxicities in Animals and Man*. pp. 91−6. (West Palm Beach: CRC Press)

Campbell, T.C. *et al.* (1980). *Feasibility of identifying adverse health effects of vitamins and essential minerals in man*. Unpublished report prepared for Food and Drug Administration, Washington, DC, by the Life Sciences Research Office. Federation of American Societies of Experimental Biology, Bethesda.

DiPalma, J.R. and Ritchie, D.M. (1977). Vitamin toxicity. *Ann. Rev. Pharmacol. Toxicol.*, **17**, 133−48

Fossati, C. (1981). Adverse reactions to vitamins. *Clin. Ther.*, **99**, 643−51

Harrison, H.E. (1978). Effect of nutrient toxicities in animals and man: vitamin D. In Rechcigl, M.Jr. (ed.) *CRC Handbook Series in Nutrition and Food. Section E: Nutritional Disorders*. Vol 1. *Effect of Nutrient Excesses and Toxicities in Animals and Man*, pp. 87−90. (West Palm Beach: CRC Press)

Haskell, B.E. (1978). Toxicity of vitamin B_6. In Rechcigl, M.Jr. (ed.), *CRC Handbook Series in Nutrition and Food. Section E: Nutritional Disorders*. Vol 1. *Effect of Nutrient Excesses and Toxicities in Animals and Man*. pp. 43−45. (West Palm Beach: CRC Press)

Hornig, D.H. and Moser, U. (1981). The safety of high vitamin C intake in man. In Counsell, J.N. and Hornig, D.H. (eds.) *Vitamin C*. pp. 225−48. (London: Applied Science Publishers)

Itokawa, Y. (1978). Effect of nutrient toxicities in animals and man: thiamine. In Rechcigl, M.Jr. (ed.) *CRC Handbook Series in Nutrition and Food. Section E: Nutritional Disorders*. Vol 1. *Effect of Nutrient Excesses and Toxicities in Animals and Man*. pp. 3−23. (West Palm Beach: CRC Press)

Jenkins, M.Y. (1978). Effect of nutrient toxicities in animals and man: Vitamin A. In Rechcigl, M.Jr. (ed.) *CRC Handbook Series in Nutrition and Food. Section E: Nutritional Disorders*. Vol 1. *Effect of Nutrient Excesses and Toxicities in Animals and Man*. pp. 73–85. (West Palm Beach: CRC Press)

Korner, W.F. and Vollm, J. (1975). New aspects of the tolerance of retinol in humans. *Int. J. Vit. Nutr. Res.*, **45**, 363–72

Marks, J. (1983). Vitamin safety. Vitamin information status paper Roche.

Preuss, H.G. (1978). Effect of nutrient toxicities in animals and man: folic acid. In Rechcigl, M.Jr. (ed.) *CRC Handbook Series in Nutrition and Food. Section E: Nutritional Disorders*. Vol 1. *Effects of Nutrient Excesses and Toxicities in Animals and Man*. pp. 61–2. (West Palm Beach: CRC Press)

Rivlin, R.S. (1978). Effect of nutrient toxicities in animals and man: Riboflavin. In Rechcigl, M.Jr. (ed.) *CRC Handbook Series in Nutrition and Food. Section E. Nutrition Disorders*. Vol 1. *Effect of Nutrient Excesses and Toxicities in Animals and Man*. pp. 25–7. (West Palm Beach: CRC Press)

Waterman, R.A. (1978). Nutrient toxicities in animals and man: niacin. In Rechcigl, M.Jr. (ed.) *CRC Handbook Series in Nutrition and Food. Section E. Nutrition Disorders*. Vol 1. *Effect of Nutrient Excesses and Toxicities in Animals and Man*. pp. 29–42. (West Palm Beach: CRC Press)

Werb, R. (1979). Vitamin A toxicity in hemodialysis patients. *Int. J. Artif. Organs*, **2**, 178–80

The specific vitamins

Vitamin A

Barker, B. (1982). Vitamin A. In Barker, B.M. and Bender, D.A. (eds.) *Vitamins and Medicine*. Vol 2, pp. 211–90. (London: Heinemann Medical)

Bates, C.J. (1983). Vitamin A in pregnancy and lactation. *Proc. Nutr. Soc.*, **42**, 65–79

Editorial. (1984). Vitamin A and cancer. *Lancet*, **2**, 325–6

Ganguly, J., Rao, M.R.S., Murthy, S.K. and Sarada, K. (1980). Systemic mode of action of vitamin A. *Vitam. Horm. (NY)*., **38**, 1–56

Glover, J. (1983). Factors affecting vitamin A transport in animals and man. *Proc. Nutr. Soc.*, **42**, 19–30

Ong, D.E. and Chytil, F. (1983). Vitamin A and cancer. *Vitam. Horm. (NY)*., **40**, 105–44

Rodriguez, M.S. and Irwin, M.I. (1972). A conspectus of research on vitamin A requirements of man. *J. Nutr.*, **102**, 909–68

Sauberlich, H.E., Hodges, R.E., Wallace, D.L., Kolder, H. et al. (1974). Vitamin A metabolism and requirements in the human studied with the use of labelled retinol. *Vitam. Horm. (NY)*., **32**, 251–75

Srikantia, S.G. (1975). Human vitamin A deficiency. In Bourne, G.H., (ed.) *World Review of Nutrition and Dietetics*. pp. 181–231. (Basel: S. Karger)

Srikantia, S.G. (1982). Vitamin A deficiency and blindness in children. In Neuberger, A. and Jukes, T.H. (eds.) *Human Nutrition: Current Issues and Controversies*. pp. 185–96. (Lancaster: MTP Press)

Vijayaraghavan, K., Pralhad Rao, N., Rameshwar Sarma, K.V., Vinodini Reddy, J. (1984). Impact of massive doses of vitamin A on incidence of nutritional blindness. *Lancet*, **2**, 149–51

WHO/USAID. (1976). Joint WHO/USAID Meeting 1976 – Vitamin A deficiency and xerophthalmia. *WHO Tech. Rep. Ser.* **590**, 17

Wolf, G. (1980). Vitamin A. In Alfin-Slater, R.B. and Kritchevsky, D. (eds.) *Nutrition and the Adult*. Vol 36. *Micronutrients*, p. 97. (New York: Academic Press)

Zile, M.H. et al. (1983). The functions of vitamin A, current concepts. *Proc. Soc. Exp. Biol. Med.*, **172**, 139–52

Vitamin D

Arnaud, S.B., Stickler, G.G. and Haworth, J.C. (1976). Serum 25-hydroxyvitamin D in infantile rickets. *Pediatrics*, **57**, 221–5

DeLuca, H.F. (1980). Vitamin D. In Alfin-Slater, R.B. and Kritchevsky, D. (eds.) *Nutrition and the Adult*. Vol 36. *Micronutrients*, p. 205. (New York: Academic Press)

DeLuca, H.F. (1982). Vitamin D. In Neuberger, A. and Jukes, T.H. (eds.) *Human Nutrition: Current Issues and Controversies*. pp. 141–83. (Lancaster: MTP Press)

DeLuca, H.F., Paaren, H.E. and Schnoes, H.K. (1979). Vitamin D and calcium metabolism. In Dewar, M.J.S., Hafner, K., Heilbronner, E., Ito, S., Lehn, J.-M., Niedenzu, K., Rees, C.W., Shafer, K., Wittig, G. and Boschke, F.L. (eds.) *Topics in Current Chemistry*. pp. 1–65. (Berlin: Springer-Verlag)

DeLuca, H.F. and Schnoes, H.K. (1979). Recent developments in the metabolism of vitamin D. In Norman, A.W., Schaefer, K., Herrath, D.V., Grigoleit, H.G. *et al.* (eds.) *Vitamin D: Basic Research and Its Clinical Application*. pp. 445–58. (Berlin: Walter de Gruyter)

Edelstein, S. (1974). Vitamin D. binding proteins. *Vitam. Horm. (NY).*, **32**, 407–28

Faccini, J.M., Exton-Smith, A.N. and Boyde, A. (1976). Disorders of bone and fractures of the femoral neck. *Lancet*, **1**, 1089–92

Fraser, D.R. (1980). Vitamin D. In Barker, B.M. and Bender, D.A. (eds.) *Vitamins in Medicine*. 2 Vols, 4th Edn., pp. 42–146. (London: Heinemann Medical)

Fraser, D.R. (1983). The physiological economy of vitamin D. *Lancet*, **1**, 969–72

Henry, H.L. (1977). Regulation of the metabolism of 25-OH-D_3: 1,25-(OH)$_2$$D_3$-parathyroid interactions. In Norman, A.W., Schaefer, K. *et al.* (eds.) *Vitamin D: Biochemical, Chemical and Clinical Aspects Related to Calcium Metabolism*. pp. 125–33 (Berlin: Walter de Gruyter)

Hunt, S.P., O'Riordan, J.L.H., Windo, J. and Truswell, A.S. (1976). Vitamin D status in different subgroups of British Asians. *Br. Med. J.*, **2**, 1351–4

Jubizw, K., Haussler, M.R., Melain, T.A. and Tolman, K.G. (1977). Plasma 1,25-dihydroxyvitamin D levels in patients receiving anticonvulsant drugs. *J. Clin. Endocrinol. Metab.*, **44**, 617–21

Lakdawalda, D.R. and Widdowson, E.M. (1977). Vitamin D in human milk. *Lancet*, **1**, 167–8

Lawson, D.E.M. and Davie, M. (1979). Aspects of the metabolism and function of vitamin D. *Vitam. Horm (NY).*, **37**, 2–68

Marx, S.J., Liberman, V.A. and Eil, C. (1983). Calciferols: actions and deficiencies. *Vitam. Horm. (NY).*, **40**, 235–308

Pierides, A.M., Ellis, H.A. and Norman, A.W. (1977). 1,25-dihydroxycholecalciferol in renal osteodystrophy. *Arch. Dis. Childh.*, **52**, 464–72

Rizvi, S.N.A. and Vaishnava, H. (1977). Occult osteomalacia in pregnant women in India. *Lancet*, **1**, 1102

Sheltawym., Newton, H. *et al.* (1984). The contribution of dietary vitamin D and sunlight to the plasma 25-hydroxyvitamin D in the elderly. *Hum. Nutr. Clin. Nutr.*, **38C**, 191–4

Vitamin E

Bauernfeind, J.C. (1980). Food sources of the tocopherols. In Machlin, L.J. (ed.) *Vitamin E.* pp. 93–135. (New York: Marcel Dekker)

Bieri, J.G. and Evarts, R.P. (1973). Tocopherols and fatty acids in American diets *J. Am. Dietet. Assoc.*, **62**, 147–1

Buri, J.C. and Farrell, P.M. (1976). Vitamin E. *Vitam. Horm. (NY).*, **34**, 31–76

Desai, J.D. (1980). Assay methods. In Machlin, L.J. (ed.) *Vitamin E.* pp. 55–98. (New York: Marcel Dekker)

Losowsky, M.S., Kelleher, J., Walker, B.E., Davies, T. and Smith, C.L. (1972). Intake and absorption of tocopherol. *Ann. NY Acad. Sci.*, **203**, 212–22

Machlin, L.J. and Brin, M. (1980). Vitamin E. In Alfin-Slater and Kritchevsky, D. (eds.) *Nutrition and the Adult. Vol 36. Micronutrients*, p. 245. (New York: Academic Press)

McCay, P.B. and King, M.M. (1980). Vitamin E: Its role as a biological free radical scavenger and its relationship to the microsomal mixed-function oxidase system. In Machlin, L.J. (ed.) *Vitamin E.* p. 79. (New York: Marcel Dekker)

Molenaar, I., Hulstaert, C.E. and Hardenk, M.J. (1980). Role in function and ultrastructure of cellular membranes. In Machlin, L.J. (ed.) *Vitamin E.* p. 117. (New York: Marcel Dekker)

Nelson, J. (1980). The pathology of vitamin E deficiency. In Machlin, L.J. (ed.) *Vitamin E.* p. 210. (New York: Marcel Dekker)

Nelson, J.S. and Fischer, V.W. (1980). Vitamin E. In Barker, B.M. and Bender, D.A. (eds.) *Vitamins in Medicine*. 2 Vols, 4th Edn., pp. 147–71. (London: Heinemann Medical)

Smith, C.L., Kelleher, J., Losowsky, M.S. and Morrish, N. (1971). The content of vitamin E in British diets. *Br. J. Nutr.*, **26**, 89–96

Thompson, J.N., Beare-Rogers, J.L., Erdody, P. and Smith, D.C. (1973). Appraisal of human vitamin E requirement based on examination of individual meals and a composite Canadian diet. *Am. J. Clin. Nutr.*, **26**, 1349–54

Witting, L.A. (1972). Recommended dietary allowance for vitamin E. *Am. J. Clin. Nutr.*, **25**, 257–61

Vitamin K

Barkhan, P. and Shearer, M.J. (1977). Metabolism of vitamin K_1 (phylloquinone) in man. *Proc. R. Soc. Med.*, **70**, 93–6

Dam, H., Sondegaerd, E. and Olson, R.E. (1982). Vitamin K. In Barker, B.M. and Bender, D.A. (eds.) *Vitamins and Medicine.* pp. 92–111. Vol 2. (London: Heinemann Medical)

Olsen, R.E. (1980). Vitamin K. In Alfin-Slater, R.B. and Kritchevsky, D. (eds.) *Nutrition and the Adult.* Vol 36. *Micronutrients*, p. 267. (New York: Academic Press)

Olson, R.E. (1974). New concepts relating to the mode of action of vitamin K. *Vitam. Horm. (NY).*, **32**, 483

Vitamin B_1

Dancy, M., Evans, G., Gaitonde, M.K. and Maxwell, J.D. (1984). Blood thiamine and thiamine phosphate ester concentrations in alcoholic and non-alcoholic liver diseases. *Br. Med. J.*, **289**, 79–82

Dreyfus, P.M. (1976). Thiamine and the nervous system: an overview, *J. Nutr. Sci. Vitaminol.*, (Suppl.), **22**, 13–16

Krampitz, L.O. (1969). Catalytic functions of thiamine diphosphate. In Snell, E.E., Boyer, P.E., Meister, A. and Sinsheimer, R.L. (eds.) *Annual Reviews of Biochemistry.* Vol. 38, pp. 123–40. (Palo Alto: Annual Reviews, Inc.)

Leder, I.G. (1975). Thiamine, biosynthesis and function. In Greenberg, D.M. (ed.) *Metabolic Pathways.* Vol. VII, 3rd Edn., pp. 57–85. (New York: Academic Press)

Sinclair, H.M. (1982). Thiamine. In Barker, B.M. and Bender, D.A. (eds.) *Vitamins & Medicine.* Vol. 2 pp. 114–67. (London: Heinemann Medical)

Vitamin B_2

Beinert, H. (1963). Electron-transferring flavoprotein. In Boyer, P.D., Lardy, H., and Myrback, K. (eds.) *The Enzymes.* Vol 7, 2nd Edn., pp. 467–76. (New York: Academic Press)

Bright, H.J. and Porter, D.J.T. (1975). Flavoprotein oxidases. In Boyer, P.D. (ed.) *The Enzymes.* Vol XII, 3rd Edn. pp. 421–505. (New York: Academic Press)

Flavell Matts, S.G. (1980). Riboflavin. In Barker, B.M. and Bender, D.A. (eds.) *Vitamins in Medicine.* 2 Vols, 4th Edn., pp. 398–438. Books (London: Heinemann Medical)

Vitamin B_6

Anonymous (1979). The vitamin B_6 requirement in oral contraceptive users. *Nutr. Rev.*, **37**, 344

Bapurao, S. and Krishnaswamy, K. (1978). Vitamin B_6 nutritional status of pellagrins and their leucine tolerance. *Am. J. Clin. Nutr.*, **31**, 819–24

Barker, B.M. and Bender, D.A. (1980). Vitamin B_6. In Barker, B.M. and Bender, D.A. (eds.) *Vitamins in Medicine.* 2 Vols, 4th Edn., pp. 348–80. (London: Heinemann Medical)

Braunstein, A.E. (1973). Amino group transfer. In Boyer, P.D. (ed.) *The Enzymes.* Vol. IX, 3rd Edn., pp. 379–481. (New York: Academic Press)

Rose, D.P. (1972). Aspects of tryptophan metabolism in health and disease: a review. *J. Clin. Pathol.*, **25**, 17–25

Niacin

Bender, D.A. (1980). Niacin. In Barker, B.M. and Bender, D.A. (eds.) *Vitamins in Medicine*. 2 Vols, 4th Edn., pp. 315–47. (London: Heinemann Medical)
Gopalan, C. and Joya Rao, K.S. (1975). Pellegra and amino acid imbalance. *Vitam. Horm. (NY).*, **33**, 505–28
Jepson, J.B. (1972). Hartnup disease. In Stanbury, J.B., Wyngaarden, J.B. and Frederickson, D.S. (eds.) *The Metabolic Basis of Inherited Disease*, 3rd Edn., pp. 1486–1503. (New York: McGraw-Hill)
Spivak, J.L. and Jackson, D.L. (1977). Pellagra: an analysis of 18 patients and a review of the literature. *Johns Hopkins Med. J.*, **140**, 295–309
Srikantia, S.G. (1982). Endemic pellagra. In Neuberger, A. and Jukes, T.H. (eds.). *Human Nutrition: Current Issues and Controversies*. pp. 209–16. (Lancaster: MTP Press)
Stratigos, J.D. and Katsambas, A.D. (1982). Pellagra: a reappraisal. *Acta Vitam. Enzymol.*, **4**, 115–21

Folic acid

Baugh, C.M. and Krumdieck, C.L. (1971). Naturally occurring folates. *Ann. NY., Acad. Sci.*, **186,** 7
Botez, M.I., Cadotte, M., Beulieu, R., Pichette, L.P. and Pison, C. (1976). Neurologic disorders responsive to folic acid therapy. *Can. Med. Assoc. J.*, **115,** 217
Brown, J.P., Scott, J.M., Foster, F.G. and Weir, D.G. (1973). Ingestion and absorption of naturally occurring pteroylmonoglutamates (folates) in man. *Gastroenterology*, **64,** 223
Chanarin, I. (1975). The folate content of foodstuffs and the availability of different folate analogues for absorption. *Getting the Most Out of Food*. No. 10, p. 41. (Burgess Hill, Sussex: Van den Berghs & Jurgens)
Chanarin, I. (1980). The folates. In Barker, B.M. and Bender, D.A. (eds.) *Vitamins in Medicine*. 2 Vols, 4th Edn., pp. 247–314. (London: Heinemann Medical)
Chanarin, I., Rothman, D., Ward, A. and Perry, J. (1968). Folate status and requirement in pregnancy. *Br. Med. J.*, **2,** 390–4
Cook, G.C., Morgan, J.O. and Hoffbrand, A.V. (1974). Impairment of folate absorption by systemic bacterial infections. *Lancet*, **2,** 1416
Erbe, R.W. (1975). Inborn errors of folate metabolism. *N. Engl. J. Med.*, **293,** 807
Goodwin, H.A. and Rosenberg, I.H. (1975). Comparative studies of the intestinal absorption of (^{3}H)pteroylmonoglutamate and (^{3}H)pteroylheptaglutamate in man. *Gastroenterology*, **69,** 364–73
Klipstein, F.A. (1972). Folate in tropical sprue. *Br. J. Haematol.*, **23,** 119
Lewis, F.B. (1974). Folate deficiency due to oral contraceptives. *Minnesota Med.*, **57,** 945
Rothenberg, S.P. and Da Costa, M. (1976). Folate binding proteins and radioassay for folate. *Clin. Haematol.*, **5,** 569
Tamura, T. and Stokstad, E.L.R. (1973). The availability of food folate in man. *Br. J. Haematol.*, **25,** 513–32

Vitamin B$_{12}$

Abeles, R.H. and Dolphin, D. (1976). The vitamin B$_{12}$ coenzyme. *Acc. Chem. Res.*, **9,** 114
Chanarin, I. (1980). The cobalamins (vitamin B$_{12}$). In Barker, B.M. and Bender, D.A. (eds.) *Vitamins in Medicine*. 2 Vols, 4th Edn., pp. 172–246. (London: Heinemann Medical)
Chanarin, I., Perry, J. and Lumb, M. (1974). The biochemical lesion in vitamin B$_{12}$ deficiency in man. *Lancet*, **1,** 1251
Dawson, D.W., Sawers, A.H. and Sharma, R.K. (1984). Malabsorption of protein bound vitamin B$_{12}$. *Br. Med. J.*, **288,** 675–8
Farquharson, J. and Adams, J.F. (1976). The forms of vitamin B$_{12}$ in foods. *Br. J. Nutr.*, **36,** 127–36
Herbert, V. (1968). Nutritional requirements for vitamin B$_{12}$ and folic acid. *Am. J. Clin. Nutr.*, **21,** 743–52
Herbert, V. and Das, K.C. (1976). The role of vitamin B$_{12}$ and folic acid in hemato- and other cell-soiesis. *Vitam. Horm. (NY).*, **34,** 2–30

Linnell, J.C. (1975). The fate of cobalamins *in vivo*. In Babior, B.M. (ed.) *Cobalamin, Biochemistry and Pathophysiology*. p. 137. (New York: Wiley Interscience)

Reynolds, E.H. (1976). Neurological aspects of folate and vitamin B_{12} metabolism. *Clin. Haematol.*, **5,** 661

Pantothenic acid

Abiko, Y. (1975). Metabolism of coenzyme A. In Greenberg, D.M. (ed.) *Metabolic Pathways*. pp. 1–25. (New York: Academic Press)

Jeffery, D.M. (1982). Panthothenic acid. In Barker, B.M. and Bender, D.A. (eds.) *Vitamins and Medicine*. Vol. 2, pp. 69–91. (London: Heinemann Medical)

Biotin

Lane, M.D. and Moss, J. (1971). The biotin-dependent enzymes. *Adv. Enzymol.*, **35,** 321

Mistry, S.P. (1980). Biotin. In Barker, B.M. and Bender D.A. (eds.) *Vitamins in Medicine*, 2 Vols, 4th Edn. pp. 381–97. (London: Heinemann Medical)

Vitamin C

Bates, C.J. (1981). The function and metabolism of vitamin C in man. In Counsell, J.N. and Hornig, D.H. (eds.) *Vitamin C – Ascorbic Acid*. pp. 1–22. (London: Applied Science Publishers)

Bender, D.A. (1982). Vitamin C. In Barker, B.M. and Bender, D.A. (eds.) *Vitamins and Medicine*. Vol. 2, pp. 1–68. (London: Heinemann Medical)

Bonjour, J.P. (1979). Vitamins and alcoholism. I. Ascorbic acid. *Int. J. Vit. Nutr. Res.*, **49,** 434

Brin, M. (1981). Marginal vitamin C deficiency and human health. In Counsell, J.N. and Hornig, D.H. (eds.) *Vitamin C – Ascorbic Acid*. pp. 359–76. (London: Applied Science Publishers)

Cooke, J.R. and Moxon, R.E.D. (1981). The detection and measurement of vitamin C. In Counsell, J.N. and Hornig, D.H., (eds.) *Vitamin C – Ascorbic Acid*. pp. 167–98. (London: Applied Science Publishers)

Hodges, R.E. (1980). Vitamin C. In Alfin-Slater, R.B. and Kritchevsky, D. (eds.) *Nutrition and the Adult*. Vol. 36. *Micronutrients*, p. 73. (New York: Academic Press)

Hodges, R.E., Hood, J., Canham, J.E., Sauberlich, H.E. *et al*. (1971). Clinical manifestations of ascorbic acid deficiency in man. *Am. J. Clin. Nutr.*, **24,** 432–43

Irvin, T.T. (1981). Vitamin C in surgical patients. In Counsell, J.N. and Hornig, D.H. (eds.) *Vitamin C – Ascorbic Acid*. pp. 283–98. (London: Applied Science Publishers)

Irwin, M.I. and Hutchins, B.K. (1976). A conspectus of research on vitamin C requirements of man. *J. Nutr.*, **106,** 823–79

Kallner, A., Hartmann, D. and Hornig, D. (1977). On the absorption of ascorbic acid in man. *Int. J. Vit. Nutr. Res.*, **47,** 383

Kallner, A. (1981). Vitamin C – man's requirement. In Counsell, J.N. and Hornig, D.H. (eds.) *Vitamin C – Ascorbic Acid*. pp. 63–73. (London: Applied Science Publishers)

Nienhuis, A.W. (1981). Vitamin C and iron. *N. Engl. J. Med.*, **304,** 170–1

Ohshima, H. and Bartsch, H. (1981). The influence of vitamin C on the *in vivo* formation of nitrosamines. In Counsell, J.N. and Hornig, D.H. (eds.) *Vitamin C – Ascorbic Acid*. pp. 215–24. (London: Applied Science Publishers)

Schorah, C.J. (1981). Vitamin C status in population groups. In Counsell, J.N. and Hornig, D.H. (eds.) *Vitamin C – Ascorbic Acid*. pp. 23–47. (London: Applied Science Publishers)

Substances chemically related to the vitamins

Bollag, W. (1983). Vitamin A and retinoids: from nutrition to pharmacotherapy in dermatology and oncology. *Lancet*, **1,** 860–3

Chan, J.C.M. and DeLuca, H.F. (1979). Calcium and parathyroid disorders in children. Chronic renal failure and treatment with calcitriol. *J. Am. Med. Assoc.*, **241,** 1242

Index

In this index the 'trivial' names of the vitamins are used, except for 'biotin', 'niacin' and 'pantothenic acid'